Random Graph Dynamics

The theory of random graphs began in the late 1950s in several papers by
Erdös and Rényi. In the late twentieth century, the notion of six degrees of
separation, meaning that any two people on the planet can be connected
by a short chain of people who know each other, inspired Strogatz and
Watts to define the small world random graph in which each site is connected to k close neighbors, but also has long-range connections. At about
the same time, it was observed in human social and sexual networks and
on the Internet that the number of neighbors of an individual or computer
has a power law distribution. This inspired Barabási and Albert to define
the preferential attachment model, which has these properties. These two
papers have led to an explosion of research. While this literature is extensive, many of the papers are based on simulations and nonrigorous
arguments. The purpose of this book is to use a wide variety of mathematical argument to obtain insights into the properties of these graphs.
A unique feature of this book is the interest in the dynamics of process
taking place on the graph in addition to their geometric properties, such
as connectedness and diameter.

Rick Durrett received his Ph.D. in operations research from Stanford
University in 1976. After nine years at UCLA and twenty-five at Cornell
University, he moved to Duke University in 2010, where he is a professor
of mathematics. He is the author of 8 books and more than 170 journal
articles on a wide variety of topics, and he has supervised more than 40
Ph.D. students. He is a member of the National Academy of Science and
the American Academy of Arts and Sciences and a Fellow of the Institute
of Mathematical Statistics.

Random Graph Dynamics

RICK DURRETT

Duke University

CAMBRIDGE
UNIVERSITY PRESS

University Printing House, Cambridge CB2 8BS, United Kingdom

One Liberty Plaza, 20th Floor, New York, NY 10006, USA

477 Williamstown Road, Port Melbourne, VIC 3207, Australia

4843/24, 2nd Floor, Ansari Road, Daryaganj, Delhi - 110002, India

79 Anson Road, #06-04/06, Singapore 079906

Cambridge University Press is part of the University of Cambridge.

It furthers the University's mission by disseminating knowledge in the pursuit of
education, learning and research at the highest international levels of excellence.

www.cambridge.org
Information on this title: www.cambridge.org/9780521150163

© Rick Durrett 2007

First published 2007
First paperback edition 2010

A catalogue record for this publication is available from the British Library

Library of Congress Cataloging in Publication data
Durrett, Richard, 1951–
Random graph dynamics / Rick Durrett.
    p. cm. – (Cambridge series in statistical and probabilistic mathematics)
Includes bibliographical references.
ISBN-13: 978-0-521-86656-9 (hardback)
ISBN-10: 0-521-86656-1 (hardback)
1. Random graphs.   I. Title.   II. Series.
QA166.17.D87   2006
511'.5–dc22   2006020925

ISBN 978-0-521-86656-9 Hardback
ISBN 978-0-521-15016-3 Paperback

Contents

v

Preface

Chapter 1 will explain what this book is about. Here I will explain why I chose to write the book, how it is written, when and where the work was done, and who helped.

Why. It would make a good story if I was inspired to write this book by an image of Paul Erdős magically appearing on a cheese quesadilla, which I later sold for thousands of dollars on eBay. However, that is not true. The three main events that led to this book were (i) the use of random graphs in the solution of a problem that was part of Nathanaël Berestycki's thesis; (ii) a talk that I heard Steve Strogatz give on the CHKNS model, which inspired me to prove some rigorous results about their model; and (iii) a book review I wrote on the books by Watts and Barabási for the *Notices of the American Math Society*.

The subject of this book was attractive for me, since many of the papers were outside the mathematics literature, so the rigorous proofs of the results were, in some cases, interesting mathematical problems. In addition, since I had worked for a number of years on the properties of stochastic spatial models on regular lattices, there was the natural question of how the behavior of these systems changed when one introduced long-range connections between individuals or considered power law degree distributions. Both of these modifications are reasonable if one considers the spread of influenza in a town where children bring the disease home from school, or the spread of sexually transmitted diseases through a population of individuals that have a widely varying number of contacts.

How. The aim of this book is to introduce the reader to the subject in the same way that a walk through Museé d'Orsay exposes the visitor to the many styles of impressionism. We will choose results to highlight the major themes, but we will not examine in detail every variation of preferential attachment that has been studied. We will concentrate on the ideas, giving the interesting parts of proofs, and referring the reader to the literature for the missing details. As Tom Liggett said after he had written his book *Interacting Particle Systems*, there is no point in having a book that is just a union of papers.

vii

Throughout we approach the subject with a probabilistic viewpoint. One pragmatic reason is that, in the absence of futuristic procedures like the one Tom Cruise's character had in *The Minority Report*, these are the only eyes through which I can view the world. For connections to computer algorithms and their analysis, you will have to ask someone who knows that story. In addition, we will emphasize topics not found in other mathematical books. I have nothing to add to the treatment of random regular graphs in Janson, Luczak, and Ruciński (2000), so I will not spend much time on this special case of random graphs with a fixed degree distribution. The classical theory of random graphs of Erdös and Rényi is covered nicely by Bollobás (2001), so I will keep treatment to the minimum necessary to prepare for more complicated examples.

Several reviewers lobbied for an introductory chapter devoted to some of the tools: branching processes, large deviations, martingales, convergence of Markov chains, almost exponentiality of waiting times, etc. Personally I do not think it is necessary (or even desirable) to read the entire *Kama Sutra* before having sex the first time, so instead I approach the book as I do my Ph.D. students' research projects. We will start looking at the subject and learn about the tools as they arise. Readers who find these interruptions distracting can note the statement and skip the proof, advice that can be applied to most of the results in the book.

When and where. The first version of these notes was written for a graduate seminar, which I gave on the topic at Cornell in the fall of 2004. On the Monday of the first full week of the semester, as I sat at my desk in Malott Hall, my back began to hurt, and I thought to myself that I had worked too hard on notes over the weekend. Two months, new desk chairs, a variety of drugs, physical therapy, and a lot of pain later, an MRI showed my problem was an infected disk. I still remember the radiologist's exciting words: "We can't let you go home until we get in touch with your doctor." Four days in the hospital and four months of sipping coffee every morning while an IV-dripped antibiotics into my arm, the bugs in me had been defeated, and I was back to almost normal. Reading papers on random graphs, figuring out proofs, and organizing the material while lying on my bed helped me to get through that ordeal.

In the summer of 2005, I revised the notes and added new material in preparation for six lectures on this topic, which I gave at the first Cornell Probability Summer School. Several more iterations of polishing followed. When my brain told me that the manuscript was in great shape, several paid reviewers showed me that there was still work to do. Finally, a month in Paris at École Normale Supérieure in February 2006 provided a pleasant setting for finishing the project. I would like to thank Jean-François LeGall for the invitation to visit Paris, radio station 92.1 FM for providing the background music while I typed in my apartment, and the restaurants on and around Rue Mouffetard for giving me something to look forward to at the end of the day.

Who. If I were Sue Grafton, the title of this book would be "G is for Random Graphs." Continuing in the tradition of the first six books, I will update the story of my family by saying that my older son David is a freshman at Ithaca College studying journalism, while Greg is a senior who has been accepted at MIT and will go there to study computer science. With 8 years of tuition, room and board, and books to pay for in the next 5 years, I desperately need you to buy this book, or better yet put $50 in an envelope and mail it to me.

In the last two decades of diapers, ear infections, special education meetings, clarinet lessons, after-school activities, weekend music events, summer internships, and driving tests, my wife Susan Myron has been the one with the more difficult job. There are no words that can adequately convey my happiness after 25 years of marriage, except that I am hoping for many more.

I would like to thank Mark Newman for his good-natured answers to several random e-mails. Postdocs Paul Jung and Lea Popovic read several of the early chapters in detail and made a number of useful suggestions. The anonymous reviewers who each read one or two chapters helped illuminate the dark corners of the manuscript and contributed some useful insights. Lauren Cowles did a wonderful job of managing the process, and the book is much better for her efforts.

As usual, I look forward to your constructive criticisms and corrections by e-mail to rtd1@cornell.edu and you can look for lists of typos, etc., on my Web page:

www.math.cornell.edu/~durrett

There you can find also copies of my recent papers, most of which concern probability problems that arise from biology.

The second printing contains a number of corrections uncovered when Norio Konno and Masato Takei translated the book into Japanese. We would also like to thank Gabriela Grüniger for helpful comments.

1

Overview

1.1 Introduction to the Introduction

The theory of random graphs began in the late 1950s in several papers by Erdős and Rényi. However, the introduction at the end of the twentieth century of the small world model of Watts and Strogatz (1998) and the preferential attachment model of Barabási and Albert (1999) have led to an explosion of research. Querying the Science Citation Index in early July 2005 produced 1154 citations for Watts and Strogatz (1998) and 964 for Barabási and Albert (1999). Survey articles of Albert and Barabási (2002), Dorogovtsev and Mendes (2002), and Newman (2003) each have hundreds of references. A book edited by Newman, Barabási, and Watts (2006) contains some of the most important papers. Books by Watts (2003) and Barabási (2002) give popular accounts of the new science of networks, which explains "how everything is connected to everything else and what it means for science, business, and everyday life."[1]

While this literature is extensive, many of the papers are outside the mathematical literature, which makes writing this book a challenge and an opportunity. A number of articles have appeared in *Nature* and *Science*. These journals with their impressive impact factors are, at least in the case of random graphs, the home of *10 second sound bite science*. An example is the claim that "the Internet is robust yet fragile. 95% of the links can be removed and the graph will stay connected. However, targeted removal of 2.3% of the hubs would disconnect the Internet."

These shocking statements grab headlines. Then long after the excitement has subsided, less visible papers show that these results aren't quite correct. When 95% of links are removed the Internet is connected, but the fraction of nodes in the giant component is 5.9×10^{-8}, so if all 6 billion people were connected initially then after the links are removed only 360 people can check their e-mail. The targeted removal result depends heavily on the fact that the degree distribution was assumed to be exactly a power law for all values of k, which forces $p_k \sim 0.832 k^{-3}$. However,

---

[1] This is the subtitle of Barabási's book.

1

if the graph is generated by the preferential attachment model with $m = 2$ then $p_k \sim 12k^{-3}$ and one must remove 33% of the hubs. See Section 4.7 for more details.

Many of the papers we cover were published in *Physical Review E*. In these we encounter the usual tension when mathematicians and physicists work on the same problems. Feynman once said "if all of mathematics disappeared it would set physics back one week." In the other direction, mathematicians complain when physicists leap over technicalities, such as throwing away terms they don't like in differential equations. They compute critical values for random graphs by asserting that cluster growth is a branching process and then calculating when the mean number of children is > 1. Mathematicians worry about justifying such approximations and spend a lot of effort coping with paranoid delusions, for example, in Section 4.2 that a sequence of numbers all of which lie between 1 and 2 might not converge.

Mathematicians cherish the rare moments where physicists' leaps of faith get them into trouble. In the current setting, physicists use the branching process picture of cluster growth when the cluster is of order n (and the approximation is not valid) to compute the average distance between points on the giant component of the random graph. As we will see, the correct way to estimate the distance from x to y is to grow the clusters until they have size $C\sqrt{n}$ and argue that they will intersect with high probability. In most cases, the two viewpoints give the same answer, but in the case of some power law graphs, the physicists' argument misses a power of 2, see Section 4.5.

While it is fun to point out physicists' errors, it is much more satisfying when we discover something that they don't know. Barbour and Reinert (2001) have shown for the small world and van der Hofstad, Hooghiemstra, and Znamenski (2005) have proved for models with a fixed degree distribution, see Theorems 5.2.1 and 3.4.1, that the fluctuations in the distance between two randomly chosen points are $O(1)$, a result that was not anticipated by simulation. We have been able to compute the critical value of the Ising model on the small world exactly, see Section 5.4, confirming the value physicists found by simulation. A third example is the Kosterlitz–Thouless transition in the CHKNS model. The five authors who introduced this model (only one of whom is a physicist) found the phenomenon by numerically solving a differential equation. Physicists Dorogovtsev, Mendes, and Samukhin (2001) demonstrated this by a detailed and semi-rigorous analysis of a generating function. However, the rigorous proof of Bollobás, Janson, and Riordan (2005), which is not difficult and given in full in Section 7.4, helps explain why this is true.

Despite remarks in the last few paragraph, our goal is not to lift ourselves up by putting other people down. As Mark Newman said in an e-mail to me "I think there's room in the world for people who have good ideas but don't have the rigor to pursue them properly – makes more for mathematicians to do." The purpose of this book is to give an exposition of results in this area and to provide proofs for some facts that had been previously demonstrated by heuristics and simulation, as well as to establish some new results. This task is interesting since it involves a wide variety of mathematics: random walks, large deviations, branching processes, branching random walks, martingales, urn schemes, and the modern theory of Markov chains that emphasizes quantitative estimates of convergence rates.

Much of this book concentrates on geometric properties of the random graphs: primarily emergence of a giant component and its small diameter. However, our main interest here is in processes taking place on these graphs, which is one of the two meanings of our title, *Random Graph Dynamics*. The other meaning is that we will be interested in graphs such as the preferential attachment model and the CHKNS model described in the final section that are grown dynamically rather than statically defined.

1.2 Erdős, Rényi, Molloy, and Reed

In the late 1950s, Erdős and Rényi introduced two random graph models. In each there are n vertices. In the first and less commonly used version, one picks m of the $n(n-1)/2$ possible edges between these vertices at random. Investigation of the properties of this model tells us what a "typical" graph with n vertices and m edges looks like. However, there is a small and annoying amount of dependence caused by picking a fixed number of edges, so here we will follow the more common approach of studying the version in which each of the $n(n-1)/2$ possible edges between these vertices are independently present with probability p. When $p = 2m/n(n-1)$, the second model is closely related to the first.

Erdős and Rényi discovered that there was a sharp threshold for the appearance of many properties. One of the first properties that was studied, and that will be the focus of much of our attention here, is the emergence of a giant component.

- If $p = c/n$ and $c < 1$ then, when n is large, most of the connected components of the graph are small, with the largest having only $O(\log n)$ vertices, where the O symbol means that there is a constant $C < \infty$ so that the probability the largest component is $\leq C \log n$ tends to 1 as $n \to \infty$.
- In contrast if $c > 1$ there is a constant $\theta(c) > 0$ so that for large n the largest component has $\sim \theta(c)n$ vertices and the second largest component is $O(\log n)$. Here $X_n \sim b_n$ means that X_n/b_n converges to 1 in probability as $n \to \infty$.

Chapter 2 is devoted to a study of this transition and properties of Erdős–Rényi random graphs below, above, and near the critical value $p = 1/n$. Much of this material is well known and can be found in considerably more detail in Bollobás' (2001) book, but the approach here is more probabilistic than combinatorial, and in any case an understanding of this material is important for tackling the more complicated graphs, we will consider later.

In the theory of random graphs, most of the answers can be guessed using the heuristic that the growth of the cluster is like that of a branching process. In *Physical Review E*, these arguments are enough to establish the result. To explain the branching process approximation for Erdős–Rényi random graphs, suppose we start with a vertex, say 1. It will be connected to a Binomial $(n - 1, c/n)$ number of neighbors, which converges to a Poisson distribution with mean c as $n \to \infty$. We consider the neighbors of 1 to be its children, the neighbors of its neighbors to be its grandchildren, and so forth. If we let Z_k be the number of vertices at distance k, then for small k, Z_k behaves like a branching process in which each individual has an independent and mean c number of children.

There are three sources of error: (i) If we have already exposed $Z_0 + \cdots + Z_k = m$ vertices then the members of the kth generation have only $n - m$ new possibilities for connections; (ii) Two or more members of the kth generation can have the same child; and (iii) Members of the branching process that have no counterpart in the growing cluster can have children. In Section 2.2, we will show that when $m = o(\sqrt{n})$, that is, $m/\sqrt{n} \to 0$, the growing cluster is equal to the branching process with high probability, and when $m = O(n^{1-\epsilon})$ with $\epsilon > 0$ the errors are of a smaller order than the size of the cluster.

When $c < 1$ the expected number of children in generation k is c^k which converges to 0 exponentially fast and the largest of the components containing the n vertices will be $O(\log n)$. When $c > 1$ there is a probability $\theta(c) > 0$ that the branching process does not die out. To construct the giant component, we argue that with probability $1 - o(n^{-1})$ two clusters that grow to size $n^{1/2+\epsilon}$ will intersect. The result about the second largest component comes from the fact with probability $1 - o(n^{-1})$ a cluster that reaches size $C \log n$ will grow to size $n^{1/2+\epsilon}$. An error term that is $o(n^{-1})$ guarantees that with high probability all clusters will do what we expect.

When $c > 1$ clusters that don't die out grow like c^k (at least as long as the branching process approximation is valid). Ignoring the parenthetical phrase we can set $c^k = n$ and solve to conclude that the giant component has "diameter" $k = \log n/(\log c)$. For a concrete example suppose $n = 6$ billion people on the planet and the mean number of neighbors $c = np = 42.62$. In this case, $\log n/(\log c) = 6$, or we have six degrees of separation between two randomly chosen individuals. We have placed diameter in quotation marks since it is commonly used in the physics literature for the distance between two randomly chosen points on the giant component. On the Erdős–Renyi random graphs the mathematically defined diameter is $\geq C \log n$ with $C > 1/\log c$, but exact asymptotics are not known, see the discussion after Theorem 2.4.2.

The first four sections of Chapter 2 are the most important for later developments. The next four can be skipped by readers eager to get to recent developments. In Section 2.5, we prove a central limit theorem for the size of the giant component. In Section 2.6, which introduces the combinatorial viewpoint, we show that away from the critical value, that is, for $p = c/n$ with $c \neq 1$, most components are trees with sizes given by the Borel–Tanner distribution. A few components, $O(1)$, have one cycle, and only the giant component is more complicated.

Section 2.7 is devoted to the critical regime $p = 1/n + \theta/n^{4/3}$, where the largest components are of order $n^{2/3}$ and there can be components more complex than unicyclic. There is a wealth of detailed information about the critical region. The classic paper by Janson, Knuth, Luczak, and Pittel (1993) alone is 126 pages. Being a probabilist, we are content to state David Aldous' (1997) result which shows that in the limit as $n \to \infty$ the growth of large components is a multiplicative coalescent.

In Section 2.8, we investigate the threshold for connectivity, that is, ALL vertices in ONE component. As Theorem 2.8.1 shows and 2.8.3 makes more precise, the Erdős–Rényi random graph becomes connected when isolated vertices disappear, so the threshold $= (\log n)/n + O(1)$. The harder, upper bound, half of this result is used in Section 4.5 for studying the diameter of random graphs with power law degree distributions.

In Chapter 3, we turn our attention to graphs with a fixed degree distribution that has finite second moment. Bollobás (1988) proved results for the interesting special case of a random r-regular graph, but Molloy and Reed (1995) were the first to construct graphs with a general distribution of degrees. Here, we will use the approach of Newman, Strogatz, and Watts (2001, 2002) to define our model. Let $d_1, \ldots d_n$ be independent and have $P(d_i = k) = p_k$. Since we want d_i to be the degree of vertex i, we condition on $E_n = \{d_1 + \cdots + d_n \text{ is even}\}$. To construct the graph now we imagine d_i half-edges attached to i, and then pair the half-edges at random. The resulting graph may have self-loops and multiple edges between points. The number is $O(1)$ so this does not bother me, but if you want a nice clean graph, you can condition on the event A_n that there are no loops or multiple edges, which has $\lim_{n \to \infty} P(A_n) > 0$.

Again, interest focuses first on the existence of a giant component, and the answer can be derived by thinking about a branching process, but the condition is not that the mean $\sum_k k p_k > 1$. If we start with a given vertex x then the number of neighbors (the first generation in the branching process) has distribution p_j. However, this is not true for the second generation. A first generation vertex with degree k is k times as likely to be chosen as one with degree 1, so the distribution of the number of children of a first generation vertex is for $k \geq 1$

$$q_{k-1} = \frac{k p_k}{\mu} \quad \text{where} \quad \mu = \sum_k k p_k$$

The $k - 1$ on the left-hand side comes from the fact that we used up one edge connecting to the vertex. Note that since we have assumed p has finite second moment, q has finite mean $\nu = \sum_k k(k-1) p_k/\mu$.

q gives the distribution of the number of children in the second and all subsequent generations so, as one might guess, the condition for the existence of a giant component is $v > 1$. The number of vertices in the kth generation grows like μv^{k-1}, so using the physicist's heuristic, the average distance between two points on the giant component is $\sim \log n/(\log v) = \log_v n$. This result is true and there is a remarkable result of van der Hofstad, Hooghiemstra, and Van Mieghem (2004a), see Theorem 3.4.1, which shows that the fluctuations around the mean are $O(1)$. Let H_n be the distance between 1 and 2 in the random graph on n vertices, and let $\bar{H}_n = (H_n | H_n < \infty)$. The Dutch trio showed that $H_n - [\log_v n]$ is $O(1)$, that is, the sequence of distributions is tight in the sense of weak convergence, and they proved a very precise result about the limiting behavior of this quantity. As far as I can tell the fact that the fluctuations are $O(1)$ was not guessed on the basis of simulations.

Section 3.3 is devoted to an

Open problem. *What is the size of the largest component when $v < 1$?*

The answer, $O(\log n)$, for Erdős–Renyi random graphs is not correct for graphs with a fixed degree distribution. For an example, suppose $p_k \sim Ck^{-\gamma}$ with $\gamma > 3$ so that the variance is finite. The degrees have $P(d_i > k) \sim Ck^{-(\gamma-1)}$ (here and in what follows C is a constant whose value is unimportant and may change from line to line). Setting $P(d_i > k) = 1/n$ and solving, we conclude that the largest of the n degrees is $O(n^{1/(\gamma-1)})$. Trivially, the largest component must be at least this large.

Conjecture. *If $p_k \sim Ck^{-\gamma}$ with $\gamma > 3$ then the largest cluster is $O(n^{1/(\gamma-1)})$.*

One significant problem in proving this is that in the second and subsequent generations the number of children has distribution $q_k \sim Ck^{-(\gamma-1)}$. One might think that this would make the largest of the n degrees $O(n^{1/(\gamma-2)})$, but this is false. The size-biased distribution q can only enhance the probability of degrees that are present in the graph, and the largest degree present is $O(n^{1/(\gamma-1)})$.

In support of the conjecture in the previous paragraph we will now describe a result of Chung and Lu (2002a, 2002b), who have introduced a variant of the Molloy and Reed model that is easier to study. Their model is specified by a collection of weights $w_1, \ldots, w_n$ that represent the expected degree sequence. The probability of an edge between i and j is $w_i w_j / \sum_k w_k$. They allow loops from i to i so that the expected degree at i is

$$\sum_j \frac{w_i w_j}{\sum_k w_k} = w_i$$

Of course, for this to make sense we need $(\max_i w_i)^2 < \sum_k w_k$.

Let $d = (1/n) \sum_k w_k$ be the average degree. As in the Molloy and Reed model, when we move to neighbors of a fixed vertex, vertices are chosen proportional to their weights, that is, i is chosen with probability $w_i / \sum_k w_k$. Thus the relevant quantity for connectedness of the graph is the second-order average degree $\bar{d} = \sum_i w_i^2 / \sum_k w_k$.

Theorem 3.3.2. *Let* $vol(S) = \sum_{i \in S} w_i$. *If* $\bar{d} < 1$ *then all components have volume at most* $A\sqrt{n}$ *with probability at least*

$$1 - \frac{d\bar{d}^2}{A^2(1 - \bar{d})}$$

Note that when $\gamma > 3$, $1/(\gamma - 1) < 1/2$ so this is consistent with the conjecture.

1.3 Six Degrees, Small Worlds

As Duncan Watts (2003) explains in his book *Six Degrees*, the inspiration for his thesis came from his father's remark that he was only six handshakes away from the president of the United States. This remark is a reference to "six degrees of separation," a phrase that you probably recognize, but what does it mean? There are a number of answers.

Answer 1. The most recent comes from the "Kevin Bacon game" that concerns the film actors graph. Two actors are connected by an edge if they appeared in the same movie. The objective is to link one actor to another by a path of the least distance. As three college students who were scheming to get on Jon Stewart's radio talk show observed, this could often be done efficiently by using Kevin Bacon as an intermediate.

This strategy leads to the concept of a Bacon number, that is, the shortest path connecting the actor to Kevin Bacon. For example, Woody Allen has a Bacon number of 2 since he was in *Sweet and Lowdown* with Sean Penn, and Sean Penn was in *Mystic River* with Kevin Bacon. The distribution of Bacon numbers given in the next table shows that most actors have a small Bacon number, with a median value of 3:

0	1	2	3	4	5	6	7	8
1	1,673	130,851	349,031	84,615	6,718	788	107	11

The average distance from Kevin Bacon for all actors is 2.94, which says that two randomly chosen actors can be linked by a path through Kevin Bacon in an average of 6 steps. Albert Barabási, who will play a prominent role in the next section, and his collaborators, computed the average distance from each person to all of the others in the film actors graph. They found that Rod Steiger with an average distance of 2.53 was the best choice of intermediate. It took them a long time to find Kevin Bacon on their list, since he was in 876th place.

Erdős Numbers. The collaboration graph of mathematics, in which two individuals are connected by an edge if they have coauthored a paper, is also a small world. The Kevin Bacon of mathematics is Paul Erdős, who wrote more than 1500 papers with more than 500 coauthors. Jerrold Grossman (2000) used 60 years of data from MathSciNet to construct a mathematical collaboration graph with 337,454 vertices (authors) and 496,489 edges. There were 84,115 isolated vertices. Discarding these gives a graph with average degree 3.92, and a giant component with 208,200 vertices with the remaining 45,139 vertices in 16,883 components. The average Erdős number is 4.7 with the largest known finite Erdős number within mathematics being 15. Based on a random sample of 66 pairs, the average distance between two individuals was 7.37. These numbers are likely to change over time. In the 1940s, 91% of mathematics papers had one author, while in the 1990s only 54% did.

Answer 2. The phrase "six degrees of separation" statement is most commonly associated with a 1967 experiment conducted by Stanley Milgram, a Harvard social psychologist, who was interested in the average distance between two people. In his study, which was first published in the popular magazine *Psychology Today* as "The Small World Problem," he gave letters to a few hundred randomly selected people in Omaha, Nebraska. The letters were to be sent toward a target person, a stockbroker in Boston, but recipients could send the letters only to someone they knew on a first-name basis. Thirty-five percent of the letters reached their destination and the median number of steps these letters took was 5.5. Rounding up gives "six degrees of separation."

   The neat story in the last paragraph becomes a little more dubious if one looks at the details. One third of the test subjects were from Boston, not Omaha, and one-half of those in Omaha were stockbrokers. A large fraction of the letters never reached their destination and were discarded from the distance computation. Of course, those that reached their destination only provide an upper bound on the distance, since there might have been better routes.

Answer 3. Though it was implicit in his work, Milgram never used the phrase "six degrees of separation." John Guare originated the term in the title of his 1990 play. In the play Ousa, musing about our interconnectedness, tells her daughter, "Everybody on the planet is separated by only six other people. Six degrees of separation. Between us and everybody else on this planet. The president of the United States. A gondolier in Venice . . . It's not just the big names. It's anyone. A native in a rain forest. A Tierra del Fuegan. An Eskimo. I am bound to everyone on this planet by a trail of six people. It is a profound thought."

Answer 4. While the Guare play may be the best known literary work with this phrase, it was not the first. It appeared in Hungarian writer Frigyes Karinthy's story Chains. "To demonstrate that people on Earth today are much closer than ever, a member of the group suggested a test. He offered a bet that we could name any person among the earth's one and a half billion inhabitants and through at most five acquaintances, one of which he knew personally, he could link to the chosen one."

Answer 5. Our final anecdote is a proof by example. A few years ago, the staff of the German newspaper *Die Zeit* accepted the challenge of trying to connect a Turkish kebab-shop owner to his favorite actor Marlon Brando. After a few months of work, they found that the kebab-shop owner had a friend living in California, who works alongside the boyfriend of a woman, who is the sorority sister of the daughter of the producer of the film *Don Juan de Marco*, in which Brando starred.

In the answers we have just given, it sometimes takes fiddling to make the answer six, but it is clear that the web of human contacts and the mathematical collaboration graph have a much smaller diameter than one would naively expect. Albert, Jeong, and Barabási (1999) and Barabási, Albert, and Jeong (2000) studied the World Wide Web graph whose vertices are documents and whose edges are links. Using complete data on the domain nd.edu at his home institution of Notre Dame, and a random sample generated by a web crawl, they estimated that the average distance between vertices scaled with the size of the graph as $0.35 + 2.06 \log n$. Plugging in their estimate of $n = 8 \times 10^8$ web pages at the time they obtained 18.59. That is, two randomly chosen web pages are on the average 19 clicks from each other. The logarithmic dependence of the distance is comforting, because it implies that "if the web grows by a 1000 per cent, web sites would still only be separated by an average of 21 clicks."

Small World Model. Erdős–Rényi graphs have small diameters, but have very few triangles, while in social networks if A and B are friends and A and C are friends, then it is fairly likely that B and C are also friends. To construct a network with small diameter and a positive density of triangles, Watts and Strogatz started from a ring lattice with n vertices and k edges per vertex, and then rewired each edge with probability p, connecting one end to a vertex chosen at random. This construction interpolates between regularity ($p = 0$) and disorder ($p = 1$).

Let $L(p)$ be the average distance between two randomly chosen vertices and define the clustering coefficient $C(p)$ to be the fraction of connections that exist between the $\binom{k}{2}$ pairs of neighbors of a site. The regular graph has $L(0) \sim n/2k$ and $C(0) \approx 3/4$ if k is large, while the disordered one has $L(1) \sim (\log n)/(\log k)$ and $C(1) \sim k/n$. Watts and Strogatz (1998), showed that $L(p)$ decreases quickly near 0, while $C(p)$ changes slowly so there is a broad interval of p over which $L(p)$ is almost as small as $L(1)$, yet $C(p)$ is far from 0. These results will be discussed in Section 5.1.

Watts and Strogatz (1998) were not the first to notice that random long distance connections could drastically reduce the diameter. Bollobás and Chung (1988) added a random matching to a ring of n vertices with nearest neighbor connections and showed that the resulting graph had diameter $\sim \log_2 n$. This graph, which we will call the *BC small world*, is not a good model of a social network because every individual has exactly three friends including one long-range acquaintance, however these weaknesses make it easier to study.

The small world is connected by definition, so the first quantity we will investigate is the average distance between two randomly chosen sites in the small world. For this problem and all of the others we will consider below, we will not rewire edges but instead consider Newman and Watts (1999) version of the model in which no edges are removed but one adds a Poisson number of shortcuts with mean $n\rho/2$ and attaches then to randomly chosen pairs of sites. This results in a Poisson mean ρ number of long distance edges per site. We will call this the *NW small world*.

Barbour and Reinert (2001) have done a rigorous analysis of the average distance between points in a continuum model in which there is a circle of circumference L and a Poisson mean $L\rho/2$ number of random chords. The chords are the shortcuts and have length 0. The first step in their analysis is to consider an upper bound model that ignores intersections of growing arcs and that assumes each arc sees independent Poisson processes of shortcut endpoints. Let $S(t)$ be size, that is, the Lebesgue measure, of the set of points within distance t of a chosen point and let $M(t)$ be the number of intervals. Under our assumptions

$$S'(t) = 2M(t)$$

while $M(t)$ is a branching process in which there are no deaths and births occur at rate 2ρ.

$M(t)$ is a Yule process run at rate 2ρ so $EM(t) = e^{2\rho t}$ and $M(t)$ has a geometric distribution

$$P(M(t) = k) = (1 - e^{-2\rho t})^{k-1} e^{-2\rho t}$$

Being a branching process $e^{-2\rho t} M(t) \to W$ almost surely. In the case of the Yule process, it is clear from the distribution of $M(t)$, that W has an exponential distribution with mean 1. Integrating gives

$$ES(t) = \int_0^t 2e^{2\rho s}\, ds = \frac{1}{\rho}(e^{2\rho t} - 1)$$

At time $t = (2\rho)^{-1}(1/2)\log(L\rho)$, $ES(t) = (L/\rho)^{1/2} - 1$. Ignoring the -1 we see that if we have two independent clusters run for this time then the expected number of connections between them is

$$\sqrt{\frac{L}{\rho}} \cdot \rho \cdot \frac{\sqrt{L/\rho}}{L} = 1$$

since the middle factor gives the expected number of shortcuts per unit distance and the last one the probability a shortcut will hit the second cluster. The precise result is:

Theorem 5.2.1. *Suppose* $L\rho \to \infty$. *Let O be a fixed point of the circle, choose P at random, and let D be the distance from O to P. Then*

$$P\left[D > \frac{1}{\rho}\left(\frac{1}{2}\log(L\rho) + x\right)\right] \to \int_0^\infty \frac{e^{-y}}{1 + 2e^{2x}y}\,dy$$

Note that the fluctuations in the distance are of order 1.

Sections 5.3, 5.4, and 5.5 are devoted to a discussion of processes taking place on the small world. We will delay discussion of these results until after we have introduced our next family of examples.

1.4 Power Laws, Preferential Attachment

One of my favorite quotes is from the 13 April 2002 issue of *The Scientist*

> What do the proteins in our bodies, the Internet, a cool collection of atoms and sexual networks have in common? One man thinks he has the answer and it is going to transform the way we view the world.

Albert-László Barabási (the man in the quote above) and Reka Albert (1999) noticed that the actor collaboration graph and the World Wide Web had degree distributions that were power laws $p_k \sim Ck^{-\gamma}$ as $k \to \infty$. Follow up work has identified a large number of examples with power law degree distributions, which are also called *scale-free random graphs*. When no reference is given, the information can be found in the survey article by Dorogovtsev and Mendes (2002). We omit biological networks (food webs, metabolic networks, and protein interaction networks) since they are much smaller and less well characterized compared to the other examples.

- By the World Wide Web, we mean the collection of web pages and the oriented links between them. Barabási and Albert (1999) found that the in-degree and out-degrees of web pages follow power laws with $\gamma_{in} = 2.1$, $\gamma_{out} = 2.7$.
- By the Internet, we mean the physically connected network of routers that move e-mail and files around the Internet. Routers are united into domains. On the interdomain level the Internet is a small network. In April 1998, when Faloutsos, Faloutsos, and Faloutsos (1999) did their study, there were 3,530 vertices, 6,432 edges and the maximum degree was 745, producing a power law with $\gamma = 2.16$. In 2000, there were about 150,000 routers connected by 200,000 links and a degree distribution that could be fit by a power law with $\gamma = 2.3$.
- The movie actor network in which two actors are connected by an edge if they have appeared in a film together has a power law degree distribution with $\gamma = 2.3$.

- The collaboration graph in a subject is a graph with an edge connecting two people if they have written a paper together. Barabási et al. (2002) studied papers in mathematics and neuroscience published in 1991–1998. The two databases that they used contained 70,901 papers with 70,975 authors, and 210,750 papers with 209,293 authors, respectively. The fitted power laws had $\gamma_M = 2.4$ and $\gamma_{NS} = 2.1$.

- Newman (2001a, 2001b) studied the collaboration network in four parts of what was then called the Los Alamos preprint archive (and is now called the arXiv). He found that the number of collaborators was better fit by a power law with an exponential cutoff $p_k = Ck^{-\tau}\exp(-k/k_c)$.

- The citation network is a directed graph with an edge from i to j if paper i cites paper j. Redner (1998) studied 783,339 papers published in 1981 in journals cataloged by the ISI and 24,296 papers published in volumes 11–50 of *Physical Review D*. The first graph had 6,716,198 links, maximum degree 8,904, and $\gamma_{in} = 2.9$. The second had 351,872 links, maximum degree 2,026, and $\gamma_{in} = 2.6$. In both cases the out degree had an exponentially decaying tail. One reason for the rapid decay of the out-degree is that many journals have a limit on the number of references.

- Liljeros et al. (2001) analyzed data gathered in a study of sexual behavior of 4,781 Swedes, and found that the number of partners per year had $\gamma_{male} = 3.3$ and $\gamma_{female} = 3.5$.

- Ebel, Mielsch, and Bornholdt (2002) studied e-mail network of Kiel University, recording the source and destination of every e-mail to or from a student account for 112 days. They found a power law for the degree distribution with $\gamma = 1.81$ and an exponential cutoff at about 100. Recently the Federal Energy Regulatory Commission has made a large e-mail data set available posting 517,341 e-mails from 151 users at Enron.

To give a mechanistic explanation for power laws Barabási and Albert (1999) introduced the preferential attachment model. For a mental image you can think of a growing World Wide Web in which new pages are constantly added and they link to existing pages with a probability proportional to their popularity. Suppose, for concreteness, that the process starts at time 1 with two vertices linked by m parallel edges. (We do this so that the total degree at any time t is $2mt$.) At every time $t \geq 2$, we add a new vertex with m edges that link the new vertex to m vertices already present in the system. To incorporate preferential attachment, we assume that the probability π_i that a new vertex will be connected to a vertex i depends on the connectivity of that vertex, so that $\pi_i = k_i / \sum_j k_j$. To be precise, when we add a new vertex we will add edges one a time, with the second and subsequent edges doing preferential attachment using the updated degrees. This scheme has the desirable property that a graph of size n for a general m can be obtained by running the $m = 1$ model for nm steps and then collapsing vertices $km, km - 1, \ldots (k-1)m + 1$ to make vertex k.

The first thing to be proved for this model, see Theorem 4.1.4, is that the fraction of vertices of degree k converges to:

$$p_k \to \frac{2m(m+1)}{k(k+1)(k+2)} \quad \text{as } n \to \infty$$

This distribution $\sim Ck^{-3}$ as $k \to \infty$, so for any value of m we always get a power of 3. Krapivsky, Redner, and Leyvraz (2000) showed that we can get other behavior for p_k by generalizing the model so that the vertices of degree k are chosen with probability proportional to $f(k)$.

- if $f(k) = k^\alpha$ with $\alpha < 1$ then $p_k \approx \mu k^{-\alpha} \exp(-ck^{1-\alpha})$
- if $f(k) = k^\alpha$ with $\alpha > 1$ then the model breaks down: there is one vertex with degree of order $\sim t$ and the other vertices have degree $O(1)$
- if $f(k) = a + k$ and $a > -1$, $p_k \sim Ck^{-(3+a)}$.

In the last case we can achieve any power in $(2, \infty)$. However, there are many other means to achieving this end. Cooper and Frieze (2003a) describe a very general model in which: old nodes sometimes generate new edges, and choices are sometimes made uniformly instead of by preferential attachment. See Section 4.2 for more details and other references.

Does preferential attachment actually happen in growing networks? Liljeros is quoted in the April 2002 of *The Scientist* as saying "Maybe people become more attractive the more partners they get." Liljeros, Edling, and Amaral (2003) are convinced of the relevance of this mechanism for sexual networks, but Jones and Handcock (2003) and others are skeptical. Jeong, Néda, and Barabási (2003) and Newman (2001c) have used collaboration databases to study the growth of degrees with time. The first paper found support for $\alpha \approx 0.8$ in the actor and neuroscience collaboration networks, while Newman finds $\alpha = 1.04 \pm 0.04$ for Medline and $\alpha = 0.89 \pm 0.09$ for the arXiv, which he argues are roughly compatible with linear preferential attachment. However, as Newman observes, a sublinear power would predict a stretched exponential distribution, which is consistent with his data.

These models and results may look new, but in reality they are quite old. If we think of a vertex i with degree $d(i)$ as $d(i)$ balls of color i, then the $m = 1$ version of the preferential attachment model is just a Polya urn scheme. We pick a ball at random from the urn, return it and a ball of the same color to the urn and add a ball of a new color. For more the connection between preferential attachment and urn schemes, see the second half of Section 4.3.

Yule (1925) used a closely related branching process model for the number of species of a given genus, which produces limiting frequencies $p_k \sim Ck^{-\gamma}$ for any $\gamma > 1$. Simon (1955) introduced a model of word usage in books, where the

$(n + 1)$th word is new with probability α or is otherwise chosen at random from the previous n words, and hence proportional to their usage. The limiting frequency of words used k times $p_k \sim Ck^{-(1+1/(1-\alpha))}$. Again this allows any power in $(2, \infty)$. For more details, see Section 4.2.

The first two sections of Chapter 4 concentrate on the fraction of vertices with a fixed degree k. In Section 4.3, we shift our attention to the other end of the spectrum and look at the growth of the degrees of a fixed vertex j. Móri (2005) has used martingales to study the case $f(k) = k + \beta$ and to show that if M_n is the maximum degree when there are n vertices, then

Theorem 4.3.2. *With probability one, $n^{-1/(2+\beta)} M_n \to \mu$.*

Since $\sum_{k=K}^{\infty} p_k \sim CK^{-(2+\beta)}$ this is the behavior we should expect by analogy with maxima of i.i.d. random variables.

Having analyzed the limiting degree distribution in preferential attachment models, we turn our attention now to the distance between two randomly chosen vertices. For simplicity, we consider the fixed degree formulation in which the graph is created in one step rather than grown. When $2 < \gamma < 3$ the size-biased distribution $q_k \sim Ck^{-(\gamma-1)}$ so the mean is infinite. Let $\alpha = \gamma - 2$. In the branching process cartoon of cluster growth, the number of vertices at distance m, Z_m grows doubly exponentially fast.

Theorem 4.5.1. $\alpha^m (\log(Z_m + 1)) \to W$.

Intuitively, the limit theorem says $\log(Z_m + 1) \approx \alpha^{-m} W$, so replacing $Z_m + 1$ by n, discarding the W and solving gives $m \sim (\log \log n)/(\log(1/\alpha))$. However, the right result which van der Hofstad, Hooghiemstra, and Znamenski (2007) have proved for the fixed degrees model is that the average distance

$$\sim 2 \cdot \frac{\log \log n}{\log(1/\alpha)}$$

To see the reason for the 2, notice that if we grow clusters from x and y until they have $\sqrt{n}$ members then each process takes time $(\log \log n)/(\log(1/\alpha))$ to reach that size. In Theorem 4.5.2, we prove the upper bound for the corresponding Chung and Lu model.

In the borderline case $\gamma = 3$, Bollobás and Riordan (2004b) have shown for the preferential attachment model, see Theorem 4.6.1, that the diameter $\sim \log n/(\log \log n)$. Chung and Lu have shown for the corresponding case of their model that the distance between two randomly chosen vertices $O(\log n/(\log \log n))$, while the diameter due to dangling ends is $O(\log n)$. To foreshadow later developments, we note that if we want degree distribution function $F(x) = P(d_i \leq x)$ then we can choose the weights in Chung and Lu's model to be

$$w_i = (1 - F)^{-1}(i/n)$$

If $1 - F(x) = Bx^{-\gamma+1}$ solving gives $w_i = (nB/i)^{1/(\gamma-1)}$. Recalling that the probability of an edge from i to j in the Chung and Lu model is $p_{i,j} = w_i w_j / \sum w_k$ and $\sum w_k \sim \mu n$ since the degree distribution has finite mean μ, we see that when $\gamma = 3$

$$p_{i,j} = c/\sqrt{ij}$$

Bollobás and Riordan (2004b) prove their result by relating edge probabilities for the preferential attachment graph to this nonhomogeneous percolation problem. This process will also make an appearance in Section 7.4 in the proof of the Kosterlitz–Thouless transition for the CHKNS model, which has connectivity probabilities $c/(i \vee j)$. A recent paper of Bollobás, Janson, and Riordan (2006), which is roughly half the length of this book, investigates inhomogeneous random graphs in great detail.

1.5 Epidemics and Percolation

The spread of epidemics on random graphs has been studied extensively. There are two extremes: in the first all individuals are susceptible and there is a probability p that an infected individual will transmit the infection to a neighbor, in the second only a fraction p of individuals are susceptible, but the disease is so contagious that if an individual gets infected all of their susceptible neighbors will become infected.

In percolation terms, the first model is bond percolation, where edges are retained with probability p and deleted with probability $1 - p$. The second is site percolation, where the randomness is applied to the sites instead of the edges. Percolation is easy to study on a random graph, since the result of retaining a fraction p of the edges or sites is another random graph. Using the branching process heuristic, percolation occurs (there will be a giant component) if and only if the mean of the associated branching process is > 1. This observation is well known in the epidemic literature, where it is phrased "the epidemic will spread if the number of secondary infections caused by an infected individual is > 1."

When the degree distribution has finite variance, the condition for a supercritical bond percolation epidemic is $E(\hat{D}(\hat{D} - 1))/E(\hat{D}) > 1$ where $\hat{D}$ is the number of edges along which the disease will be transmitted, see Section 3.5. Newman (2002) was the first to do this calculation for random graphs with a fixed degree distribution, but incorrectly assumed that the transmission events for different edges are independent, which is false when the duration of the infectious period is random. While the random graph setting for epidemics is new, the associated supercriticality condition is not. May and Anderson (1988) showed that for the transmission of AIDS and other diseases where there is great heterogeneity in the number of secondary infections, k, the basic reproductive number $R_0 = \rho_0(1 + C_V^2)$ where $\rho_0 = <k>$ is the average number of secondary infections, $C_V = (<k^2> / <k>^2) - 1$ is the coefficient of variation of the connectivity distribution, and $<X>$ is physicist's notation for expected value EX.

As noted in the previous section, many networks have power law degree distributions with power $2 < \gamma < 3$. In this case the size-biased distribution q has infinite mean. Thus for any $p > 0$ the mean number of secondary contacts is > 1 and the critical value for percolation $p_c = 0$. This "surprising result" has generated a lot of press since it implies that "within the observed topology of the Internet and the www, viruses can spread even when the infection probabilities are vanishingly small."

This quote is from Lloyd and May's (2001) discussion in *Science* of Pastor-Satorras and Vespignani (2001a). This dire prediction applies not only to computers but also to sexually transmitted diseases. "Sexual partnership networks are often extremely heterogeneous because a few individuals (such as prostitutes) have very high numbers of partners. Pastor-Satorras and Vespignani's results may be of relevance in this context. This study highlights the potential importance of studies on communication and other networks, especially those with scale-free and small world properties, for those seeking to manage epidemics within human and animal populations." Fortunately, for the people of Sweden, $\gamma_{male} = 3.3$ and $\gamma_{female} = 3.5$, so sexually transmitted diseases have a positive epidemic threshold.

Dezső and Barabási (2002) continue this theme in their work: "From a theoretical perspective viruses spreading on a scale free network appear unstoppable. The question is, can we take advantage of the increased knowledge accumulated in the past few years about the network topology to understand the conditions in which one can successfully eradicate the viruses?" The solution they propose is obvious. The vanishing threshold is a consequence of the nodes of high degree, so curing the "hubs" is a cost-effective method for combating the epidemic. As Liljeros et al. (2001) say in the subhead of their paper: "promiscuous individuals are the vulnerable nodes to target in safe-sex campaigns." For more on virus control strategies for technological networks, see Balthrop, Forrest, Newman, and Williamson (2004). The SARS epidemic with its superspreaders is another situation where the highly variable number of transmissions per individual calls for us to rethink our approaches to preventing the spread of disease, see for example, Lloyd-Smith, Schreiber, Kopp and Getz (2005).

One of the most cited properties of scale-free networks, which is related to our discussion of epidemics, is that they are "robust to random damage but vulnerable to malicious attack." Albert, Jeong, and Barabási (2000) performed simulation studies on the result of attacks on a map of the Internet consisting of 6,209 vertices and 24,401 links. Their simulations and some approximate calculations suggested that 95% of the links can be removed and the graph will stay connected. Callaway,

Newman, Strogatz, and Watts (2000) modeled intentional damage as removal of the vertices with degrees $k > k_0$, where k_0 is chosen so that the desired fraction of vertices f is eliminated. They computed threshold values for the distribution $p_k = k^{-\gamma}/\zeta(\gamma)$ when $\gamma = 2.4, 2.7, 3.0$. Here ζ is Riemann's function, which in this context plays the mundane role of giving the correct normalization to produce a probability distribution. The values of f_c in the three cases are 0.023, 0.010, and 0.002, so using the first figure for the Internet, the targeted destruction of 2.3% of the hubs would disconnect the Internet.

The results in the last paragraph are shocking, which is why they attracted headlines. However, as we mentioned earlier, one must be cautious in interpreting them. Bollobás and Riordan (2004c) and Riordan (2004) have done a rigorous analysis of percolation on the Barabási–Albert preferential attachment graph, which has $\beta = 3$. In the case $m = 1$ the world is a tree and destroying any positive fraction of the edges disconnects it.

Theorem 4.7.3. *Let $m \geq 2$ be fixed. For $0 < p \leq 1$ there is a constant c_m and a function*

$$\lambda_m(p) = \exp\left(-\frac{c_m(1 + o(1))}{p}\right)$$

so that with probability $1 - o(1)$ the size of largest component is $(\lambda_m(p) + o(1))n$ and the second largest is $o(n)$.

In words, if p is small then the giant component is tiny, and it is unlikely you will be able to access the Internet from your house. In the concrete case considered above $c = 1/\zeta(3)$. Using $\zeta(3) = 1.202057$ and setting $p = 0.05$ gives the result quoted in the first section of this introduction that the fraction of nodes in the giant component in this situation is 5.9×10^{-8}.

Bollobás and Riordan (2004c) have also done a rigorous analysis of intentional damage for the preferential attachment model, which they define as removal of the first nf nodes, which are the ones likely to have the largest degrees.

Theorem 4.7.4. *Let $m \geq 2$ and $0 < f < 1$ be constant. If $f \geq (m-1)/(m+1)$ then with probability $1 - o(1)$ the largest component is $o(n)$. If $f < (m-1)/(m+1)$ then there is a constant $\theta(f)$ so that with probability $1 - o(1)$ the largest component is $\sim \theta(f)n$, and the second largest is $o(n)$.*

It is difficult to compare this with the conclusions of Callaway, Newman, Strogatz, and Watts (2000) since for any m in the preferential attachment model we have $p_k \sim 2m(m+1)k^{-3}$ as $k \to \infty$. However, the reader should note that even when $m = 2$, one can remove 1/3 of the nodes.

One of the reasons why Callaway, Newman, Strogatz, and Watts (2000) get such small number is that, as Aiello, Chung, and Lu (2000, 2001) have shown, graphs with degree distribution $p_k = k^{-\gamma}/\zeta(\gamma)$ have no giant component for $\gamma > 3.479$. Thus the fragility is an artifact of assuming that there is an exact power law, while in reality the actual answer for graphs with $p_k \sim Ck^{-\gamma}$ depends on the value of C as well. This is just one of many criticisms of the claim that the Internet is "robust yet fragile." Doyle et al. (2005) examine in detail how the scale-free depiction compares with the real Internet.

Percolation on the small world is studied in Section 5.3. Those results are the key to the ones in the next section.

1.6 Potts Models and the Contact Process

In the Potts model, each vertex is assigned a spin $\sigma(x)$ which may take one of q values. Given a finite graph G with vertices V and edges E, for example, the small world, the energy of a configuration is

$$H(\sigma) = 2 \sum_{x,y \in V, x \sim y} 1\{\sigma(x) \neq \sigma(y)\}$$

where $x \sim y$ means x is adjacent to y. Configurations are assigned probabilities proportional to $\exp(-\beta H(\sigma))$, where β is a variable inversely proportional to temperature. We define a probability measure on $\{1, 2, \ldots q\}^V$ by

$$\nu(\sigma) = Z^{-1} \exp(-\beta H(\sigma))$$

where Z is a normalizing constant that makes the $\nu(\sigma)$ sum to 1. When $q = 2$ this is the Ising model, though in that case it is customary to replace $\{1, 2\}$ by $\{-1, 1\}$, and write the energy as

$$H_2(\sigma) = - \sum_{x,y \in V, x \sim y} \sigma(x)\sigma(y)$$

This leads to the same definition of ν since every pair with $\sigma(x) \neq \sigma(y)$ increases H_2 by 2 from its minimum value in which all the spins are equal, so $H - H_2$ is constant and after normalization the measures are equal.

To study the Potts model on the small world, we will use the random–cluster model of Fortuin and Kasteleyn. This is a $\{0, 1\}$-valued process η on the edges E of the graph:

$$\mu(\eta) = Z^{-1} \left\{ \prod_{e \in E} p^{\eta(e)}(1 - p)^{1-\eta(e)} \right\} q^{\chi(\eta)}$$

where $\chi(\eta)$ is the number of connected components of η when we interpret 1-bonds as occupied and 0-bonds as vacant and Z is another normalizing constant.

Having introduced the model with a general q, we will now restrict our attention to the Ising model with $q = 2$. By using some comparison arguments we are able to show that on a general graph the Ising model has long-range order for $\beta > \beta_I$ where $\tanh(\beta_I) = p_c$, the threshold for percolation in the model with independent bonds, see Theorem 5.4.5. To explain the significance of this equation, consider the Ising model on a tree with forward branching number $b \geq 2$. The critical value for the onset of "spontaneous magnetization" has $\tanh(\beta_c) = 1/b$. This means that when $\beta > \beta_c$ if we impose $+1$ boundary conditions at sites a distance n from the root and let $n \to \infty$ then in the resulting limit spins $\sigma(x)$ have positive expected value. When $0 \leq \beta \leq \beta_c$ there is a unique limiting Gibbs state independent of the boundary conditions, see for example, Preston (1974).

This connection allows us to show in Section 5.4 that for BC small world, which looks locally like a tree of degree 3, the Ising model critical value is

$$\beta_I = \tanh^{-1}(1/2) = 0.5493$$

$1/\beta_I = 1.820$ agrees with the critical value of the temperature found from simulations of Hong, Kim, and Choi (2002), but physicists seem unaware of this simple exact result. Using results of Lyons (1989, 1990) who defined a branching number for trees that are not regular, we are able to extend the argument to the nearest neighbor NW small world, which locally like a two type branching process.

In making the connection with percolation, we have implicitly been considering the SIR (susceptible–infected–removed) epidemic model in which sites, after being infected, become removed from further possible infection. This is the situation for many diseases, such as measles, and would seem to be reasonable for computers whose anti-virus software has been updated to recognize the virus. However, Pastor-Satorras and Vespignani (2001a, 2001b, 2002) and others have also considered the SIS (susceptible–infected–susceptible) in which sites that have been cured of the infection are susceptible to reinfection. We formulate the model in continuous time with infected sites becoming healthy (and again susceptible) at rate 1, while an infected site infects each of its susceptible neighbors at rate λ. In the probability literature, this SIS model is called Harris' (1974) contact process. There it usually takes place on a regular lattice like $\mathbb{Z}^2$ and is more often thought of as a model for the spread of a plant species.

The possibility of reinfection in the SIS model allows for an endemic equilibrium in which the disease persists infecting a positive fraction of the population. Since the graph is finite the infection will eventually die out, but as we will see later, there is a critical value λ_c of the infection rate the disease persists for an extremely long time. Pastor-Satorras and Vespignani have made an extensive study of this model using mean-field methods. To explain what this means, let $\rho_k(t)$ denote the fraction of vertices of degree k that are infected at time t, and $\theta(\lambda)$ be the probability that a given link points to an infected site. If we make the mean-field assumption that there are no correlations then

$$\frac{d}{dt}\rho_k(t) = -\rho_k(t) + \lambda k[1 - \rho_k(t)]\theta(\lambda)$$

Analysis of this equation suggests the following conjectures about the SIS model on power law graph with degree distribution $p_k \sim C k^{-\gamma}$.

- If $\gamma \leq 3$ then $\lambda_c = 0$
- If $3 < \gamma < 4$, $\lambda_c > 0$ but $\theta(\lambda) \sim C(\lambda - \lambda_c)^{1/(\gamma-3)}$ as $\lambda \downarrow \lambda_c$
- If $\gamma > 4$ then $\lambda_c > 0$ and $\theta(\lambda) \sim C(\lambda - \lambda_c)$ as $\lambda \downarrow \lambda_c$.

The second and third claims are interesting open problems. Berger, Borgs, Chayes, and Saberi (2005) have considered the contact process on the Barbási–Albert preferential attachment graph. They have shown that $\lambda_c = 0$ and proved some interesting results about the probability the process will survive from a randomly chosen site. The proof of $\lambda_c = 0$ is very easy and is based on the following:

Lemma 4.8.2. *Let G be a star graph with center 0 and leaves $1, 2, \ldots k$. Let A_t be the set of vertices infected in the contact process at time t when $A_0 = \{0\}$. If $k\lambda^2 \to \infty$ then*

$$P(A_{\exp(k\lambda^2/10)} \neq \emptyset) \to 1$$

The largest degree in the preferential attachment graph is $O(n^{1/2})$ so if $\lambda > 0$ is fixed the process will survive for time at least $\exp(cn^{1/2})$.

At this point we have considered the Ising model on the small world, and the contact process on graphs with a power law degree distribution. The other two combinations have also been studied. Durrett and Jung (2005) have considered the contact process on a generalization of the BC small world and showed that like the contact process on trees the system has two phase transitions, see Section 5.5 for details.

Dorogovtsev, Goltsev, and Mendes (2002) have studied the Ising model on power law graphs. Their calculations suggest that $\beta_c = 0$ for $\gamma \leq 3$, the spontaneous magnetization $M(\beta) \sim (\beta - \beta_c)^{1/(\gamma-3)}$ for $3 < \gamma < 5$ while for $\gamma > 5$, $M(\beta) \sim (\beta - \beta_c)^{1/2}$. A rigorous proof of the results for critical exponents seems difficult, but can one use the connection between the Ising model and percolation to show $\beta_c = 0$ for $\gamma \leq 3$?

1.7 Random Walks and Voter Models

There have been quite a few papers written about the properties of random walks on small world networks studying the probability the walker is back where it started after n steps, the average number of sites visited, etc. See for example, Monasson (1999), Jespersen, Sokolov, and Blumen (2000), Lahtinen, Kertesz, and Kaski

(2001), Pandit and Amritkar (2001), Almaas, Kulkarni, and Stroud (2003), and Noh and Reiger (2004). In most cases the authors have concentrated on the situation in which the density of shortcuts p is small, and shown that for small times $t \ll \xi^2$ with $\xi = 1/p$ the behavior is like a random walk in one dimension, at intermediate times the behavior is like a random walk on a tree, and at large times the walker realizes it is on a finite set.

Here, we will concentrate instead on the rate of convergence to equilibrium for random walks. Let $K(x, y)$ be the transition kernel of the lazy walk that stays put with probability 1/2 and otherwise jumps to a randomly chosen neighbor. The laziness gets rid of problems with periodicity and negative eigenvalues. Let $\pi(x)$ be its stationary distribution, and define $Q(x, y) = \pi(x)K(x, y)$ to be the flow from x to y in equilibrium. Our walks satisfy the detailed balance condition, that is, $Q(x, y) = Q(y, x)$.

In most cases we will bound the rate of convergence to equilibrium by considering the conductance

$$h = \min_{\pi(S) \le 1/2} \frac{Q(S, S^c)}{\pi(S)}$$

where $Q(S, S^c) = \sum_{x \in S, y \in S^c} Q(x, y)$. If all of the vertices have the same degree d and we let $e(S, S^c)$ be the number of edges between S and S^c, $h = \iota/2d$ where

$$\iota = \min_{|S| \le n/2} \frac{e(S, S^c)}{|S|}$$

is the *edge isoperimetric constant*.

Cheeger's inequality and standard results about Markov chains, see Sections 6.1 and 6.2, imply that if h is bounded away from 0 as the size of the graph n tends to ∞, then convergence to equilibrium takes time that is $O(\log n)$. This result takes care of most of our examples. Bollobás (1988) estimated the isoperimetric constant for random regular graphs, the special case of a fixed degree distribution with $p_r = 1$ for some $r \ge 3$. In Section 6.3, we will prove a more general result with a worse constant due to Gkantsis, Mihail, and Saberi (2003).

Theorem 6.3.2. *Consider a random graph with a fixed degree distribution in which the minimum degree is $r \ge 3$. There is a constant $\alpha_0 > 0$ so that $h \ge \alpha_0$.*

In Sections 6.4 and 6.5, we will show that the same conclusion holds for Barabási and Albert's preferential attachment graph, a result of Mihail, Papadimitriou, and Saberi (2004), and for connected Erdős–Rényi random graphs $ER(n, (c \log n)/n)$ with $c > 1$, a result of Cooper and Frieze (2003b).

It is easy to show that the random walk on the BC small world in which each vertex has degree 3, mixes in time $O(\log n)$. In contrast the random walk on the NW small world mixes in time at least $O(\log^2 n)$ and at most $O(\log^3 n)$. The lower bound is easy to see and applies to any graph with a fixed degree distribution with $p_2 > 0$. There are paths of length $O(\log n)$ in which each vertex has degree 2. The time to escape from this interval starting from the middle is $O(\log^2 n)$ which gives a lower bound of $O(\log^2 n)$ on the mixing time. The upper bound comes from showing that the conductance $h \geq C/\log n$, which translates into a bound of order $\log^3 n$. We believe that the lower bound is the right order of magnitude. In Section 6.7, we prove this is correct for a graph with a fixed degree distribution in which $p_2 + p_3 = 1$.

The voter model is a very simple model for the spread of an opinion. On any of our random graphs it can be defined as follows. Each site x has an opinion $\xi_t(x)$ and at the times of a rate 1 Poisson process decides to change its opinion. To do this it picks a neighbor at random and adopts the opinion of that neighbor. If you don't like this simple minded sociological interpretation, you can think instead of this as a spatial version of the Moran model of population genetics.

To analyze the voter model we use a "dual process" $\zeta_s^{x,t}$ that works backwards in time to determine the source of the opinion at x at time t and jumps if voter at $\zeta_s^{x,t}$ at time $t - s$ imitated one of its neighbors. The genealogy of one opinion is a random walk. If we consider several at once we get a coalescing random walk since $\zeta_s^{x,t} = \zeta_s^{t,y}$ implies that the two processes will agree at all later times.

If we pick the starting points x and y according to the stationary distribution π for the random walk and let T_A be the time at which they first hit, Proposition 23 of Aldous and Fill (2003) implies

$$\sup_t |P_\pi(T_A > t) - \exp(-t/E_\pi T_A)| \leq \tau_2/E_\pi T_A$$

where τ_2 is the relaxation time, which they define (see p. 19) to be 1 over the spectral gap. In many of our examples $\tau_2 \leq C \log^2 n$ and as we will see $E_\pi T_A \sim cn$ so the hitting time is approximately exponential. To be precise, we will show this in Section 6.9 for the BC and NW small worlds, fixed degree distributions with finite variance, and connected Erdős–Rényi random graphs.

Holley and Liggett (1975) showed that on the d-dimensional lattice $\mathbb{Z}^d$, if $d \leq 2$ the voter model approaches complete consensus, that is, $P(\xi_t(x) = \xi_t(y)) \to 1$, while if $d \geq 3$ and we start from product measure with density p (i.e., we assign opinions 1 and 0 independently to sites with probabilities p and $1 - p$) then as $t \to \infty$, ξ_t^p converges in distribution to ξ_∞^p, a one parameter family of stationary distributions.

On a finite set the voter model will eventually reach an absorbing state in which all voters have the same opinion. Cox (1989) studied the voter model on a finite torus $(Z \bmod N)^d$ and showed that if $p \in (0, 1)$ then the time to reach consensus τ_N satisfies $\tau_N = O(s_N)$ where $s_N = N^2$ in $d = 1$, $s_N = N^2 \log N$ in $d = 2$, and $s_N = N^d$ in $d \geq 3$. Our results for the voter model on the BC or NW small worlds show that while the world starts out one dimensional, the long-range connections make the behavior like that of a voter model in $d \geq 3$, where the time to reach the absorbing state is proportional to the volume of the system. Before reaching the absorbing state the voter model settles into a quasi-stationary distribution, which is like the equilibrium state in dimensions $d \geq 3$. Castelano, Vilone, and Vespignani (2003) arrived at these conclusions on the basis of simulation.

While the voter model we have studied is natural, there is another one with nicer properties. Consider the voter model defined by picking an edge at random from the graph, flipping a coin to decide on an orientation (x, y), and then telling the voter at y to imitate the voter at x. The random walk in this version of the voter model has a uniform stationary distribution and in the words of Suchecki, Eguuíluz, and Miguel (2005): "conservation of the global magnetization." In terms more familiar to probabilists, the number of voters with a given opinion is a time change of simple random walk and hence is a martingale. If we consider the biased voter model in which changes from 0 to 1 are always accepted but changes from 1 to 0 occur with probability $\lambda < 1$, then the last argument shows that the fixation probability for a single 1 introduced in a sea of 0's does not depend on the structure of the graph, the small world version of a result of Maruyama (1970) and Slatkin (1981). Because of this property, Lieberman, Hauert, and Nowak (2005), who studied evolutionary dynamics on general graphs, call the random walk *isothermal*.

1.8 CHKNS Model

Inspired by Barabási and Albert (1999), Callaway, Hopcroft, Kleinberg, Newman, and Strogatz (2001) introduced the following simple version of a randomly grown graph. Start with $G_1 = \{1\}$ with no edges. At each time $n \geq 2$, we add one vertex and with probability δ add one edge between two randomly chosen vertices. Note that the newly added vertex is not necessarily an endpoint of the added edge and when n is large, it is likely not to be.

In the original CHKNS model, which we will call model #0, the number of edges was 1 with probability δ, and 0 otherwise. To obtain a model that we can analyze rigorously, we will study the situation in which a Poisson mean δ number of vertices are added at each step. We prefer this version since, in the Poisson case, if we let $A_{i,j,k}$ be the event no (i, j) edge is added at time k then $P(A_{i,j,k}) = \exp\left(-\delta/\binom{k}{2}\right)$ for $i < j \leq k$ and these events are independent.

$$P(\cap_{k=j}^n A_{i,j,k}) = \prod_{k=j}^{n} \exp\left(-\frac{2\delta}{k(k-1)}\right)$$
$$= \exp\left(-2\delta\left(\frac{1}{j-1} - \frac{1}{n}\right)\right) \geq 1 - 2\delta\left(\frac{1}{j-1} - \frac{1}{n}\right) \qquad \#1$$

The last formula is somewhat ugly, so we will also consider two approximations

$$\approx 1 - 2\delta \left(\frac{1}{j} - \frac{1}{n} \right) \qquad\qquad \text{#2}$$

$$\approx 1 - \frac{2\delta}{j} \qquad\qquad \text{#3}$$

The approximation that leads to #3 is not as innocent as it looks. If we let $\mathcal{E}_n$ be the number of edges at time n then using the definition of the model $E\mathcal{E}_n \sim \delta n$ in models #1 and #2 but $E\mathcal{E}_n \sim 2\delta n$ in model #3. Despite this, it turns out that models #1, #2, and #3 have the same qualitative behavior, so in the long run we will concentrate on #3.

The first task is to calculate the critical value for the existence of a giant component. CHKNS showed that the generating function $g(x)$ of the size of the component containing a randomly chosen site satisfied

$$g'(x) = \frac{1}{2\delta x} \cdot \frac{x - g(x)}{1 - g(x)}$$

and used this to conclude that if $g(1) = 1$ then the mean cluster size $g'(1) = (1 - \sqrt{1 - 8\delta})/4\delta$. Since this quantity becomes complex for $\delta > 1/8$ they concluded $\delta_c = 1/8$. See Section 7.1 for a more complete description of their argument. One may quibble with the proof but the answer is right. As we will prove in Section 7.2, in models #1, #2, or #3 the critical value $\delta_c = 1/8$.

In contrast to the situation with ordinary percolation on the square lattice where Kesten (1980) proved the physicists' answer was correct nearly 20 years after they had guessed it, this time the rigorous answer predates the question by more than 10 years. We begin by describing earlier work on the random graph model on $\{1, 2, 3, \ldots\}$ with $p_{i,j} = \lambda/(i \vee j)$. Kalikow and Weiss (1988) showed that the probability G is connected (ALL vertices in ONE component) is either 0 or 1, and that $1/4 \leq \lambda_c \leq 1$. They conjectured $\lambda_c = 1$ but Shepp (1989) proved $\lambda_c = 1/4$. To connect with the answer $\delta_c = 1/8$, note that $\lambda = 2\delta$. Durrett and Kesten (1990) proved a result for a general class of $p_{i,j} = h(i, j)$ that are homogeneous of degree -1, that is, $h(ci, cj) = c^{-1}h(i, j)$. It is their methods that we will use to prove the result.

To investigate the size of the giant component, CHKNS integrated the differential equation for the generating function g near $\delta = 1/8$. Letting $S(\delta) = 1 - g(1)$ the fraction of vertices in the infinite component they plotted $\log(-\log S)$ versus $\log(\delta - 1/8)$ and concluded that

$$S(\delta) \sim \exp(-\alpha(\delta - 1/8)^{-\beta})$$

where $\alpha = 1.132 \pm 0.008$ and $\beta = 0.499 \pm 0.001$. Based on this they conjectured that $\beta = 1/2$. Note that, in contrast to the many examples we have seen previously where $S(\delta) \sim C(\delta - \delta_c)$ as $\delta \downarrow \delta_c$, the size of the giant component is infinitely differentiable at the critical value. In the language of physics we have a Kosterlitz–Thouless transition. If you are Russian you add Berezinskii's name at the beginning of the name of the transition.

Inspired by CHKNS' conjecture Dorogovstev, Mendes, and Samukhin (2001) computed that as $\delta \downarrow 1/8$,

$$S \equiv 1 - g(1) \approx c \, \exp(-\pi/\sqrt{8\delta - 1})$$

To compare with the numerical result we note that $\pi/\sqrt{8} = 1.1107$. To derive their formula Dorogovstev, Mendes, and Samukhin (2001) change variables $u(\xi) = 1 - g(1 - \xi)$ in the differential equation for g to get

$$u'(\xi) = \frac{1}{2\delta(1 - \xi)} \cdot \frac{u(\xi) - \xi}{u(\xi)}$$

They discard the $1 - \xi$ in the denominator (without any justification or apparent guilt at doing so), solve the differential equation explicitly and then do some asymptotic analysis of the generating function, which one can probably make rigorous. The real mystery is why can you drop the $1 - \xi$?

Again one may not believe the proof, but the result is correct. Bollobás, Janson, and Riordan (2005) have shown, see Theorem 7.4.1, that if $\eta > 0$ then

$$S(\delta) \leq \exp(-(1 - \eta)/\sqrt{8\delta - 1})$$

when $\delta - \delta_c > 0$ is small, and they have proved a similar lower bound. Their proof relates the percolation process in the random graph in which i is connected to j with probability $c/(i \vee j)$ to the one in which the probability is $c/\sqrt{ij}$. The latter process played a role in the analysis of the diameter of the preferential attachment model in Section 4.6.

Recently Riordan (2004) has shown

$$S(\delta) = \exp\left(-\frac{\pi(1 + o(1))}{\sqrt{8\delta - 1}}\right)$$

In addition to the striking behavior of the size of the giant component, the behavior of the cluster size distribution is interesting at the critical point and in the subcritical regime. As we show in Section 7.3 for models #0 or #1, at $\delta = 1/8$ the probability a randomly chosen site belongs to a cluster of size k, $b_k \sim 1/(k \log k)^2$, when $\delta < 1/8$

$$b_k \sim C_\delta k^{-2/(1-\sqrt{1-8\delta})}$$

Our results on the probability of $i \to j$, that is, a path from i to j, are not as good. We are able to show, see Section 7.5 for model #3, that for $\delta = 1/8$, and $1 \leq i < j \leq n$,

$$P(i \to j) \leq \frac{3}{8}\Gamma_{i,j}^n \quad \text{where} \quad \Gamma_{i,j}^n = \frac{(\log i + 2)(\log n - \log j + 2)}{(\log n + 4)}$$

Overview

and that this implies

$$\frac{1}{n} \sum_{i=1}^{n} E|\mathcal{C}_i| \le 6$$

so the mean cluster size is finite at the critical value. However, we are not able to prove that $P(i \to j) \ge c\Gamma_{i,j}$ when i is close to 1 and j is close to n. In the subcritical regime, we prove in Section 7.3 for model #3 that if $i < j$ then

$$P(i \to j) \le \frac{c}{2ri^{1/2-r} j^{1/2+r}}$$

where $r = \sqrt{1 - 8\delta}/2$. This upper bound is an important ingredient in the proof of the Bollobás, Janson, and Riordan (2005) result in Section 7.4, but in view of our difficulties when $\delta = 1/8$, we have not investigated lower bounds.

2

Erdős–Rényi Random Graphs

In this chapter we will introduce and study the random graph model introduced by Erdős and Rényi in the late 1950s. This example has been extensively studied and a very nice account of many of the results can be found in the classic book of Bollobás (2001), so here we will give a brief account of the main results on the emergence of a giant component, in order to prepare for the analysis of more complicated examples. In contrast to other treatments, we mainly rely on methods from probability and stochastic processes rather than combinatorics.

To define the model, we begin with the set of vertices $V = \{1, 2, \ldots n\}$. For $1 \leq x < y \leq n$ let $\eta_{x,y}$ be independent $= 1$ with probability p and 0 otherwise. Let $\eta_{y,x} = \eta_{x,y}$. If $\eta_{x,y} = 1$ there is an edge from x to y. Here, we will be primarily concerned with situation $p = \lambda/n$ and in particular with showing that when $\lambda < 1$ all of the components are small, with the largest $O(\log n)$, while for $\lambda > 1$ there is a giant component with $\sim g(\lambda)n$ vertices. The intuition behind this result is that a site has a Binomial$(n - 1, \lambda/n)$ number of neighbors, which has mean $\approx \lambda$. Suppose that we start with $I_0 = 1$, and for $t \geq 1$ let I_t be the set of vertices not in $\cup_{s=0}^{t-1} I_s$ that are connected to some site in I_{t-1}. Then when t is not too large the number of points in I_t, $Z_t = |I_t|$, is approximately a branching process in which each individual in generation t has an average of λ children. If $\lambda < 1$ the branching process dies out quickly and all components are small. When $\lambda > 1$, the branching process survives with probability $g(\lambda)$, and all sites with surviving branching processes combine to make the giant component.

2.1 Branching Processes

In this section we define branching processes and gives their basic properties. Since n will be the number of vertices in our graph, we will use t and s for our discrete time parameter. Let $\xi_i^t, i, t \geq 0$, be i.i.d. nonnegative integer-valued random variables. Define a sequence Z_t, $t \geq 0$ by $Z_0 = 1$ and

$$
Z_{t+1} = \begin{cases} \xi_1^{t+1} + \cdots + \xi_{Z_t}^{t+1} & \text{if } Z_t > 0 \\ 0 & \text{if } Z_t = 0 \end{cases}
$$

Z_t is called a *Galton–Watson process*. The idea behind the definition is that Z_t is the number of people in the tth generation, and each member of the tth generation gives birth independently to an identically distributed number of children. $p_k = P(\xi_i^t = k)$ is called the *offspring distribution*.

Lemma 2.1.1. *Let* $\mathcal{F}_t = \sigma(\xi_i^s : i \geq 1, 1 \leq s \leq t)$ *and* $\mu = E\xi_i^t \in (0, \infty)$. *Then* Z_t/μ^t *is a martingale w.r.t.* $\mathcal{F}_t$.

Proof. Clearly, Z_t is measurable with respect to $\mathcal{F}_t$, or $Z_t \in \mathcal{F}_t$. Recall, see Exercise 1.1 in Chapter 4 of Durrett (2004), that if $X = Y$ on $B \in \mathcal{F}$ then $E(X|\mathcal{F}) = E(Y|\mathcal{F})$ on B. On $\{Z_t = k\}$,

$$
E(Z_{t+1}|\mathcal{F}_t) = E(\xi_1^{t+1} + \cdots + \xi_k^{t+1}|\mathcal{F}_t) = k\mu = \mu Z_t
$$

Dividing both sides by μ^{t+1} now gives the desired result. ∎

Theorem 2.1.2. *If* $\mu < 1$ *then* $Z_t = 0$ *for all t sufficiently large.*

Proof. $E(Z_t/\mu^t) = E(Z_0) = 1$, so $E(Z_t) = \mu^t$. Now $Z_t \geq 1$ on $\{Z_t > 0\}$ so

$$
P(Z_t > 0) \leq E(Z_t; Z_t > 0) = E(Z_t) = \mu^t \to 0
$$

exponentially fast if $\mu < 1$. ∎

The last answer should be intuitive. If each individual on the average gives birth to less than one child, the species will die out. The next result shows that after we exclude the trivial case in which each individual has exactly one child, the same result holds when $\mu = 1$.

Theorem 2.1.3. *If* $\mu = 1$ *and* $P(\xi_i^t = 1) < 1$ *then* $Z_t = 0$ *for all t sufficiently large.*

Proof. When $\mu = 1$, Z_t is itself a nonnegative martingale, so the martingale convergence theorem, (2.11) in Chapter 4 of Durrett (2004), implies that Z_t converges to an a.s. finite limit Z_∞. Since Z_t is integer valued, we must have $Z_t = Z_\infty$ for large t. If $P(\xi_i^t = 1) < 1$ and $k > 0$ then $P(Z_t = k$ for all $t \geq T) = 0$ for any T, so we must have $Z_\infty \equiv 0$. ∎

Theorem 2.1.4. *If* $\mu > 1$ *then* $P(Z_t > 0$ *for all* $t) > 0$.

Proof. For $\theta \in [0, 1]$, let $\phi(\theta) = \sum_{k \geq 0} p_k \theta^k$ where $p_k = P(\xi_i^t = k)$. ϕ is the *generating function* for the offspring distribution p_k. Differentiating gives for $\theta < 1$

$$\phi'(\theta) = \sum_{k=1}^{\infty} k p_k \theta^{k-1} \geq 0$$

$$\phi''(\theta) = \sum_{k=2}^{\infty} k(k-1) p_k \theta^{k-2} \geq 0$$

So ϕ is increasing and convex, and $\lim_{\theta \uparrow 1} \phi'(\theta) = \sum_{k=1}^{\infty} k p_k = \mu$. Our interest in ϕ stems from the following facts.

(a) *If* $\theta_t = P(Z_t = 0)$ *then* $\theta_t = \sum_{k=0}^{\infty} p_k \theta_{t-1}^k = \phi(\theta_{t-1})$

Proof of (a). If $Z_1 = k$, an event with probability p_k, then $Z_t = 0$ if and only if all k families die out in the remaining $t - 1$ units of time, independent events with probability θ_{t-1}^k. Summing over the disjoint possibilities for each k gives the desired result. ∎

(b) *If* $\phi'(1) = \mu > 1$ *there is a unique* $\rho < 1$ *so that* $\phi(\rho) = \rho$.

Proof of (b). $\phi(0) \geq 0$, $\phi(1) = 1$, and $\phi'(1) > 1$, so $\phi(1 - \epsilon) < 1 - \epsilon$ for small ϵ. The last two observations imply the existence of a fixed point. To see it is unique, observe that $\mu > 1$ implies $p_k > 0$ for some $k > 1$, so $\phi''(\theta) > 0$ for $\theta > 0$. Since ϕ is strictly convex, it follows that if $\rho < 1$ is the smallest fixed point, then $\phi(x) < x$ for $x \in (\rho, 1)$. ∎

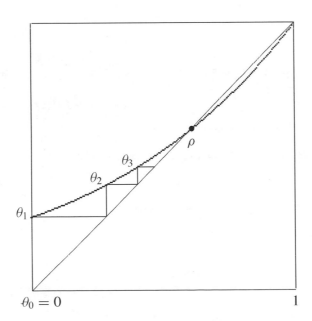

Generating function for Poisson mean $\lambda = 1.25$.

(c) *As* $t \uparrow \infty$, $\theta_t \uparrow \rho$.

Proof of (c). $\theta_0 = 0$, $\phi(\rho) = \rho$, and ϕ is increasing, so induction implies θ_t is increasing and $\theta_t \leq \rho$. Let $\theta_\infty = \lim \theta_t$. Taking limits in $\theta_t = \phi(\theta_{t-1})$, we see $\theta_\infty = \phi(\theta_\infty)$. Since $\theta_\infty \leq \rho$, it follows that $\theta_\infty = \rho$.   ∎

Combining (a)–(c) shows $P(Z_t = 0 \text{ for some } t) = \lim_{t \to \infty} \theta_t = \rho < 1$.   ∎

Example. Consider the Poisson distribution with mean λ, that is,

$$P(\xi = k) = e^{-\lambda} \frac{\lambda^k}{k!}$$

In this case $\phi(s) = \sum_{k=0}^{\infty} e^{-\lambda} s^k \lambda^k / k! = \exp(\lambda(s-1))$ so the fixed point equation is

$$\rho = \exp(\lambda(\rho - 1)) \tag{2.1.1}$$

Theorem 2.1.4 shows that when $\mu > 1$, the limit of Z_t/μ^t has a chance of being nonzero. The best result on this question is due to Kesten and Stigum.

Theorem 2.1.5. *Suppose* $\mu > 1$. $W = \lim Z_t/\mu^t$ *is not* $\equiv 0$ *if and only if* $\sum p_k k \log k < \infty$.

For a proof, see Athreya and Ney (1972, pp. 24–29). We will now prove the following simpler result.

Theorem 2.1.6. *If* $\sum_k p_k > 1$ *and* $\sum k^2 p_k < \infty$ *then* $W = \lim Z_t/\mu^t$ *is not* $\equiv 0$.

Proof. Let $\sigma^2 = \text{var}(\xi_i^t)$. Let $X_t = Z_t/\mu^t$. Writing $X_t = X_{t-1} + (X_t - X_{t-1})$.

$$E(X_t^2|\mathcal{F}_{t-1}) = X_{t-1}^2 + 2X_{t-1}E(X_t - X_{t-1}|\mathcal{F}_{t-1}) + E((X_t - X_{t-1})^2|\mathcal{F}_{t-1})$$
$$= X_{t-1}^2 + E((X_t - X_{t-1})^2|\mathcal{F}_{t-1}) \tag{2.1.2}$$

since X_t is a martingale. To compute the second term, we observe

$$E((X_t - X_{t-1})^2|\mathcal{F}_{t-1}) = E((Z_t/\mu^t - Z_{t-1}/\mu^{t-1})^2|\mathcal{F}_{t-1})$$
$$= \mu^{-2t} E((Z_t - \mu Z_{t-1})^2|\mathcal{F}_{t-1}) \tag{2.1.3}$$

On $\{Z_{t-1} = k\}$,

$$E((Z_t - \mu Z_{t-1})^2|\mathcal{F}_{t-1}) = E\left(\left(\sum_{i=1}^{k} \xi_i^t - \mu k\right)^2 \middle| \mathcal{F}_{t-1}\right) = k\sigma^2 = Z_{t-1}\sigma^2$$

Combining the last three equations gives

$$EX_t^2 = EX_{t-1}^2 + E(Z_{t-1}\sigma^2/\mu^{2t}) = EX_{t-1}^2 + \sigma^2/\mu^{t+1}$$

since $E(Z_{t-1}/\mu^{t-1}) = EZ_0 = 1$. Now $EX_0^2 = 1$, so $EX_1^2 = 1 + \sigma^2/\mu^2$, and induction gives

$$EX_t^2 = 1 + \sigma^2 \sum_{s=2}^{t+1} \mu^{-s} \qquad (2.1.4)$$

This shows $\sup_t EX_t^2 < \infty$, so by the L^2 convergence theorem for martingales, (4.5) in Chapter 4 of Durrett (2004), $X_t \to W$ in L^2, and hence $EX_t \to EW$. $EX_t = 1$ for all t, so $EW = 1$ and W is not $\equiv 0$. ∎

Our next result shows that when W is not $\equiv 0$ it is positive on the set where the branching process does not die out.

Theorem 2.1.7. *If $P(W = 0) < 1$ then $\{W > 0\} = \{Z_t > 0 \text{ for all } n\}$, that is, the symmetric difference of the two sets has probability 0.*

Proof. Let $\rho = P(W = 0)$. In order for Z_t/μ^t to converge to 0 this must also hold for the branching process started by each of the children in the first generation. Breaking things down according to the number of children in the first generation

$$\rho = \sum_{k=0}^{\infty} p_k \rho^k = \phi(\rho)$$

so $\rho < 1$ is a fixed point of the generating function and hence $\rho = P(Z_t = 0$ for some $t)$. Clearly, $\{W > 0\} \subset \{Z_t > 0 \text{ for all } t\}$. Since the two sets have the same probability $P(\{Z_t > 0 \text{ for all } t\} - \{W > 0\}) = 0$, which is the desired result. ∎

The limit theorems above describe the growth of the process when it does not die out. Our next question is: what happens in a supercritical branching process when it dies out?

Theorem 2.1.8. *A supercritical branching process conditioned to become extinct is a subcritical branching process. If the original offspring distribution is Poisson(λ) with $\lambda > 1$ then the conditioned one is Poisson($\lambda\rho$) where ρ is the extinction probability.*

Proof. Let $T_0 = \inf\{t : Z_t = 0\}$ and consider $\bar{Z}_t = (Z_t | T_0 < \infty)$. To check the Markov property for $\bar{Z}_t$ note that the Markov property for Z_t implies:

$$P(Z_{t+1} = z_{t+1}, T_0 < \infty | Z_t = z_t, \dots Z_0 = z_0) = P(Z_{t+1} = z_{t+1}, T_0 < \infty | Z_t = z_t)$$

To compute the transition probability for $\bar{Z}_t$, observe that if ρ is the extinction probability then $P_x(T_0 < \infty) = \rho^x$. Let $p(x, y)$ be the transition probability for Z_t. Note that the Markov property implies

$$\bar{p}(x, y) = \frac{P_x(Z_1 = y, T_0 < \infty)}{P_x(T_0 < \infty)} = \frac{P_x(Z_1 = y)P_y(T_0 < \infty)}{P_x(T_0 < \infty)} = \frac{p(x, y)\rho^y}{\rho^x}$$

Taking $x = 1$ and computing the generating function

$$\sum_{y=0}^{\infty} \bar{p}(1, y)\theta^y = \rho^{-1} \sum_{y=0}^{\infty} p(1, y)(\theta\rho)^y = \rho^{-1}\phi(\theta\rho)$$

where $p_y = p(1, y)$ is the offspring distribution.

$\bar{p}_y = \bar{p}(1, y)$ is the distribution of the size of the family of an individual, conditioned on the branching process dying out. If we start with x individuals then in Z_n each gives rise to an independent family. In $\bar{Z}_n$ each family must die out, so $\bar{Z}_n$ is a branching process with offspring distribution $\bar{p}(1, y)$. To prove this formally observe that

$$p(x, y) = \sum_{j_1,\ldots,j_x \geq 0, j_1 + \cdots + j_x = y} p_{j_1} \cdots p_{j_x}$$

Writing $\sum_*$ as shorthand for the sum in the last display

$$\frac{p(x, y)\rho^y}{\rho^x} = \sum_* \frac{p_{j_1}\rho^{j_1}}{\rho} \cdots \frac{p_{j_x}\rho^{j_x}}{\rho} = \sum_* \bar{p}_{j_1} \cdots \bar{p}_{j_x}$$

In the case of the Poisson distribution $\phi(s) = \exp(\lambda(s - 1))$ so if $\lambda > 1$, using the fixed point equation (2.1.1)

$$\frac{\phi(s\rho)}{\rho} = \frac{\exp(\lambda(s\rho - 1))}{\exp(\lambda(\rho - 1))} = \exp(\lambda\rho(s - 1))$$

which completes the proof.   ∎

Geometrically, a supercritical branching process conditioned to die out is a branching process with a generating function that is obtained by taking the graph of ϕ over $[0, \rho]$ and rescaling to make the domain and range $[0, 1]$. In the next result we take the graph of ϕ over $[\rho, 1]$ and rescale to make the domain and range $[0, 1]$.

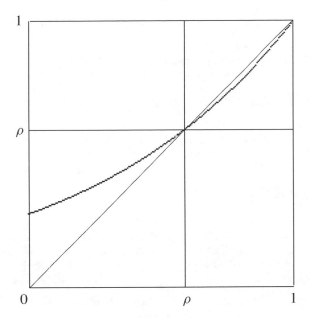

Theorem 2.1.9. *Consider a supercritical branching process with offspring distribution p_k and generating function ϕ. If we condition on nonextinction and look only at the individuals that have an infinite line of descent then the number of individuals in generation t, $\tilde{Z}_t$ is a branching process with offspring generating function*

$$\tilde{\phi}(\theta) = \frac{\phi((1-\rho)\theta + \rho) - \rho}{1 - \rho}$$

where ρ is the extinction probability, that is the smallest solution of $\phi(\rho) = \rho$ in $[0, 1]$.

Proof. There is nothing to prove if $\rho = 0$ so suppose $0 < \rho < 1$. If $Z_0 = 1$ and we condition on survival of the branching process, then the number of individuals in the first generation who have an infinite line of descent has distribution

$$\tilde{p}_j = \frac{1}{1-\rho} \sum_{k=j}^{\infty} p_k \binom{k}{j} (1-\rho)^j \rho^{k-j}$$

Multiplying by θ^j, summing, and interchanging the order of summation

$$\sum_{j=1}^{\infty} \tilde{p}_j \theta^j = \frac{1}{1-\rho} \sum_{j=1}^{\infty} \sum_{k=j}^{\infty} p_k \binom{k}{j} (1-\rho)^j \rho^{k-j} \theta^j$$

$$= \frac{1}{1-\rho} \sum_{k=1}^{\infty} p_k \sum_{j=1}^{k} \binom{k}{j} (1-\rho)^j \theta^j \rho^{k-j}$$

Using the binomial theorem and noticing that the $j = 0$ term is missing the above

$$= \frac{1}{1 - \rho} \sum_{k=1}^{\infty} p_k \{((1 - \rho)\theta + \rho)^k - \rho^k\}$$

We can add the $k = 0$ term to the sum since its value is 0. Having done this the result is

$$\frac{\phi((1 - \rho)\theta + \rho) - \phi(\rho)}{1 - \rho}$$

Since $\phi(\rho) = \rho$ the result follows.   ∎

2.2 Cluster Growth as an Epidemic

In this section we use branching process results to study the growth of the connected component, or cluster, containing 1, a process which is the same as a discrete time epidemic. To begin the construction, we let $S_0 = \{2, 3, \ldots, n\}$, $I_0 = \{1\}$, and $R_0 = \emptyset$. The letters are motivated by the epidemic interpretation of the growing cluster. S_t are the susceptibles, I_t are infected, and R_t are removed. In graph terms, we have already examined the connections of all sites in R_t, I_t are the sites to be investigated on this turn, and S_t are unexplored. These sets evolve as follows:

$$R_{t+1} = R_t \cup I_t$$
$$I_{t+1} = \{y \in S_t : \eta_{x,y} = 1 \text{ for some } x \in I_t\}$$
$$S_{t+1} = S_t - I_{t+1} \tag{2.2.1}$$

The cluster containing 1, $\mathcal{C}_1 = \cup_{t=0}^{\infty} I_t$.

Kendall (1956) was the first to suggest a branching process approximation for epidemics. To define a comparison branching process we introduce a new independent set of variables $\zeta_{x,y}^t$, $t \geq 1$, $1 \leq x, y \leq n$ that are independent, $= 1$ with probability λ/n, and 0 otherwise. Let $Z_0 = 1$, $S_t^c = \{1, 2, \ldots, n\} - S_t$ and

$$Z_{t+1} = \sum_{x \in I_t, y \in S_t} \eta_{x,y} + \sum_{x \in I_t} \sum_{y \in S_t^c} \zeta_{x,y}^t + \sum_{x=n+1}^{n+Z_t-|I_t|} \sum_{y=1}^{n} \zeta_{x,y}^t \tag{2.2.2}$$

The third term represents children of individuals in the branching process that are not in I_t. The second term, which we will denote by B_{t+1}, is the set of extra births in the branching process due to the fact that $|S_t| < n$. As for the first term,

$$C_{t+1} = \sum_{x \in I_t, y \in S_t} \eta_{x,y} - |I_{t+1}| \geq 0$$

is the number of collisions, that is, the number of births that occur in the branching process but are not matched by an increase in the cluster size. It is immediate from the construction that Z_t is a branching process with offspring distribution Binomial $(n, \lambda/n)$ and

$$Z_t \geq |I_t|$$

This is enough to take care of the case $\lambda < 1$. $EZ_t = \lambda^t$, so the mean cluster size

$$E|\mathcal{C}_1| = E\left(\sum_{t=0}^{\infty} |I_t|\right) \leq \sum_{t=0}^{\infty} \lambda^t = \frac{1}{1-\lambda} < \infty \qquad (2.2.3)$$

To study the case $\lambda > 1$ (and to show that the last result is asymptotically sharp) we need to bound the difference between Z_t and $|I_t|$. Let $Y_0 = 0$ and for $t \geq 0$ let

$$Y_{t+1} = \sum_{s=0}^{t} Z_s = Y_t + Z_t \geq |R_{t+1}|$$

Let $\mathcal{F}_t$ be the σ-field generated by S_s, I_s, R_s for $s \leq t$ and $\zeta_{x,y}^s$ with $s \leq t$.

$$E(B_{t+1}|\mathcal{F}_t) = \frac{\lambda}{n}|I_t|(|I_t| + |R_t|) \leq \frac{\lambda}{n}Z_t(Z_t + Y_t) = \frac{\lambda}{n}Z_t Y_{t+1} \qquad (2.2.4)$$

To bound the collision term we observe that

$$C_{t+1} \leq |\{(x, x', y) : x, x' \in I_t, y \in S_t, x < x', \eta_{x,y} = \eta_{x',y} = 1\}|$$

so using $|S_t| \leq n$ and $|I_t| \leq Z_t$ we have

$$E(C_{t+1}|\mathcal{F}_t) \leq \left(\frac{\lambda}{n}\right)^2 |I_t|^2 |S_t| \leq \frac{\lambda^2}{n} Z_t^2 \qquad (2.2.5)$$

To estimate the third term in (2.2.2), we call the $C_s + B_s$ individuals added at time s *immigrants*, and let $A_{s,t}$ be the number of children at time $t \geq s$ of immigrants at time s, with $A_{s,s} = B_s + C_s$. The third error term is $\sum_{s=1}^{t} A_{s,t+1}$. Clearly

$$E(A_{s,t}) = \lambda^{t-s} E(B_s + C_s) \qquad (2.2.6)$$

The next result shows that in the subcritical regime the cluster size distribution is very close to the total progeny in the branching process when n is large.

Theorem 2.2.1. *If $\lambda < 1$, $\sum_{t=1}^{\infty} E(Z_t - |I_t|) \leq \frac{C}{n}$.*

Here and in what follows C is a constant whose value is unimportant and may change from line to line.

Proof. Since $Z_t - |I_t| = \sum_{s=1}^{t} A_{s,t}$, interchanging the order of summation and using (2.2.6) gives

$$\sum_{t=1}^{\infty} E(Z_t - |I_t|) = E\left(\sum_{s=1}^{\infty} \sum_{t=s}^{\infty} A_{s,t}\right) = \frac{1}{1-\lambda} E \sum_{s=1}^{\infty} (B_s + C_s) \qquad (2.2.7)$$

Using (2.2.4), (2.2.5), and recalling $Y_{s+1} = \sum_{r=0}^{s} Z_r = Y_s + Z_s$ the above is

$$E \sum_{s=1}^{\infty} (B_s + C_s) \le \frac{\lambda}{n} E \left(\sum_{s=1}^{\infty} \sum_{r=0}^{s-1} Z_r Z_s \right) + \frac{\lambda + \lambda^2}{n} E \sum_{s=0}^{\infty} Z_s^2 \qquad (2.2.8)$$

If $r < s$ then $E(Z_r Z_s | \mathcal{F}_r) = Z_r E(Z_s | \mathcal{F}_r) = \lambda^{s-r} Z_r^2$, so

$$E(Z_r Z_s) = \lambda^{s-r} E Z_r^2 \qquad (2.2.9)$$

and interchanging the order of summation

$$E \left(\sum_{r=0}^{\infty} \sum_{s=r+1}^{\infty} Z_r Z_s \right) = \frac{\lambda}{1-\lambda} E \sum_{r=0}^{\infty} Z_r^2$$

Combining the last three displays we see that to complete the proof it is enough to show that $E \sum_{s=0}^{\infty} Z_s^2 \le C$. To do this we compute the second moments recursively. Adding and subtracting $\lambda Z_{s-1} = E(Z_s | \mathcal{F}_{s-1})$ inside the square and noting that the crossproduct term vanishes

$$E(Z_s^2 | \mathcal{F}_{s-1}) = \lambda^2 Z_{s-1}^2 + E((Z_s - \lambda Z_{s-1})^2 | \mathcal{F}_{s-1})$$
$$= \lambda^2 Z_{s-1}^2 + \sigma^2 Z_{s-1}$$

since conditional on $\mathcal{F}_{s-1}$, Z_s is a sum of Z_{s-1} random variables with mean λ and variance $\sigma^2 = n(\lambda/n)(1 - \lambda/n) \le \lambda$. $E Z_{s-1} = \lambda^{s-1}$ so

$$E Z_s^2 \le \lambda^s + \lambda^2 E Z_{s-1}^2$$

Iterating we have

$$E Z_s^2 \le \lambda^s + \lambda^{s+1} + \lambda^4 E Z_{s-2}^2$$
$$\le \lambda^s + \lambda^{s+1} + \lambda^{s+2} + \lambda^6 E Z_{s-3}^2$$
$$\le \sum_{r=s}^{\infty} \lambda^r = \lambda^s / (1-\lambda)$$

which completes the proof. ∎

Theorem 2.2.2. *If* $\lambda > 1$, $E(Z_t - |I_t|) \le \frac{C}{n} \lambda^{2t+2}$.

Proof. Using (2.2.8) with (2.2.9) as in the previous proof

$$E(B_s + C_s) \le \frac{\lambda}{n} E \left(\sum_{r=0}^{s-2} Z_r Z_{s-1} \right) + \frac{\lambda + \lambda^2}{n} E Z_{s-1}^2 \le \frac{\lambda + \lambda^2}{n} \left(\sum_{r=0}^{s} \lambda^{s-r} E Z_r^2 \right)$$

Using (2.1.4), $EZ_r^2 \leq C\lambda^{2r}$, so $E(B_s + C_s) \leq C\lambda^{2s+1}/n$, and it follows from (2.2.7) that

$$E(Z_t - |I_t|) = E\left(\sum_{s=1}^{t} A_{s,t}\right) \leq \frac{C}{n}\lambda^{2t+2}$$

and the proof is complete. ∎

When $t = a\log n/(\log\lambda)$, Theorem 2.2.2 becomes

$$E(Z_t - |I_t|) \leq Cn^{2a-1}$$

If $a < 1/2$ then the right-hand side tends to 0 and $Z_t = |I_t|$ with a probability that tends to 1 as $n \to \infty$. $EZ_t = \lambda^t = n^a$ so if $a < 1$, $E(Z_t - |I_t|) = o(EZ_t)$. In words, the cluster is almost the same size as the branching process at times when the expected size is $o(n)$.

2.3 Cluster Growth as a Random Walk

Although the connection with branching processes is intuitive, it is more convenient technically to expose the cluster one site at a time to obtain something that can be approximated by a random walk. In this section we will introduce that connection and use it to prove the existence of a giant component with $\sim\theta(\lambda)n$ vertices when $p = \lambda/n$ with $\lambda > 1$.

Since we have an emotional attachment to using S_t for a random walk, we will change the previous notation and let $R_0 = \emptyset$, $U_0 = \{2, 3, \ldots, n\}$, and $A_0 = \{1\}$. R_t is the set of removed sites, U_t are the unexplored sites and A_t is the set of active sites. At time $\tau = \inf\{t : A_t = \emptyset\}$ the process stops. If $A_t \neq \emptyset$, pick i_t from A_t according to some rule that is measurable with respect to $\mathcal{A}_t = \sigma(A_0, \ldots A_t)$ and let

$$\begin{aligned}
R_{t+1} &= R_t \cup \{i_t\} \\
A_{t+1} &= A_t - \{i_t\} \cup \{y \in U_t : \eta_{i_t,y} = 1\} \\
U_{t+1} &= U_t - \{y \in U_t : \eta_{i_t,y} = 1\}
\end{aligned} \tag{2.3.1}$$

This time $|R_t| = t$ for $t \leq \tau$, so the cluster size is τ.

Upper Bound for $\lambda < 1$. To define a comparison random walk, we introduce a new independent set of variables ζ_y^t, $t \geq 1$, $y \leq n$ that are independent, $= 1$ with probability λ/n, and 0 otherwise. Let $S_0 = 1$ and for $t \geq 0$, let $U_t^c = \{1, 2, \ldots, n\} - U_t$

$$S_{t+1} = S_t - 1 + \begin{cases} \sum_{y \in U_t} \eta_{i_t,y} + \sum_{y \in U_t^c} \zeta_y^t & \text{if } A_t \neq \emptyset \\ \sum_{y=1}^{n} \zeta_y^t & \text{if } A_t = \emptyset \end{cases}$$

S_t is a random walk with $S_t \geq |A_t|$ if $t \leq \tau$, so if $T = \inf\{t : S_t = 0\}$ then $\tau \leq T$.

The increments X_i of the random walk are $-1 + \text{Binomial}(n, \lambda/n)$. If $\lambda < 1$ stopping the martingale $S_t - (\lambda - 1)t$ at the bounded stopping time $T \wedge t$ gives

$$ES_{T \wedge t} - (\lambda - 1)E(T \wedge t) = ES_0 = 1$$

Since $ES_{T \wedge t} \geq 0$, it follows that $E(T \wedge t) \leq 1/(1 - \lambda)$. Letting $t \to \infty$ we have $ET \leq 1/(1 - \lambda)$. Having verified that $ET < \infty$ we can now use Wald's equation, see for example, (1.6) in Chapter 3 of Durrett (2004), to conclude $E(S_T - S_0) = (\lambda - 1)ET$ and hence

$$ET = 1/(1 - \lambda) \tag{2.3.2}$$

We can get a much better result by using the moment generating function.

Theorem 2.3.1. *Suppose* $\lambda < 1$ *and let* $\alpha = \lambda - 1 - \log(\lambda) > 0$. *If* $a > 1/\alpha$ *then*

$$P\left(\max_{1 \leq x \leq n} |\mathcal{C}_x| \geq a \log n\right) \to 0$$

Remark. This bound is very accurate. Corollary 5.11 of Bollobás (2001) shows that the size of the largest component is asymptotically

$$\frac{1}{\alpha}\left(\log n - \frac{5}{2} \log \log n\right) + O(1)$$

Proof. We begin by computing the moment generating function:

$$\begin{aligned}
\phi(\theta) = E \exp(\theta X_i) &= e^{-\theta}(1 - \lambda/n + (\lambda/n)e^{\theta})^n \\
&\leq \exp(-\theta + \lambda(e^{\theta} - 1)) = \psi(\theta)
\end{aligned} \tag{2.3.3}$$

since $1 + x \leq e^x$. Note that the right-hand side is the moment generating function of $-1 + \text{Poisson}$ mean λ. $\psi'(0) = \lambda - 1$ so if $\lambda < 1$ then $\psi(\theta) < 1$ when $\theta > 0$ is small. To optimize we set the derivative

$$\frac{d}{d\theta}(-\theta + \lambda(e^{\theta} - 1)) = -1 + \lambda e^{\theta} = 0$$

when $\theta = \theta_1 = -\log \lambda$. At this point $\psi(\theta_1) = \exp(\log(\lambda) + 1 - \lambda) \equiv e^{-\alpha} < 1$. $\exp(\theta_1 S_t)/\phi(\theta_1)^t$ is a nonnegative martingale, so using the optional stopping theorem for the nonnegative supermartingale $M_t = \exp(\theta_1 S_t)/\psi(\theta_1)^t$, see for example, (7.6) in Chapter 4 of Durrett (2004)

$$1/\lambda = e^{\theta_1} \geq E(\psi(\theta_1)^{-T}) = E(e^{\alpha T})$$

so using Chebyshev's inequality

$$P(T \geq k) \leq e^{-k\alpha}/\lambda \tag{2.3.4}$$

Letting $\mathcal{C}_x$ denote the cluster containing x, noting that $T \geq |\mathcal{C}_x|$ in distribution, and taking $k = (1 + \epsilon)(\log n)/\alpha$

$$P(|\mathcal{C}_x| \geq (1+\epsilon)(\log n)/\alpha) \leq n^{-(1+\epsilon)}/\lambda$$

from which the desired result follows. ∎

Lower Bound for $\lambda > 1$. To get a lower bound on the growth of the cluster let $\hat{U}_t^\delta$ consists of the $(1-\delta)n$ smallest members of $\hat{U}_t$. As long as $\hat{A}_t \neq \emptyset$ and $\hat{U}_t^\delta \geq (1-\delta)n$ which corresponds to $|\hat{A}_t| + t \leq n\delta$, we can define

$$\hat{R}_{t+1} = \hat{R}_t \cup \{j_t\}$$
$$\hat{A}_{t+1} = \hat{A}_t - \{j_t\} \cup \{y \in \hat{U}_t^\delta : \eta_{j_t,y} = 1\}$$
$$\hat{U}_{t+1} = \hat{U}_t - \{y \in \hat{U}_t^\delta : \eta_{j_t,y} = 1\} \qquad (2.3.5)$$

where $j_t = \min \hat{A}_t$. It is easy to see that if we take $i_t = j_t$ in (2.3.1) then $|A_t| \geq |\hat{A}_t|$. To define a comparison random walk, we let $W_0 = 1$ and for

$$T_W = \inf\{s : W_s = 0, \text{ or } W_s + s \geq n\delta\}$$

define

$$W_{t+1} = W_t - 1 + \begin{cases} \sum_{y \in \hat{U}_t^\delta} \eta_{j_t,y} & \text{if } t < T_W \\ \sum_{y=1}^{n(1-\delta)} \zeta_y^t & \text{if } t \geq T_W \end{cases}$$

It is easy to see that for $t \leq T_W$, $|\hat{A}_t| = W_t$ so $\tau \geq T_W$.

We will use the comparison random walk to prove

Theorem 2.3.2. *Suppose* $\lambda > 1$. *There is a constant* β *so that with probability* $\to 1$, *there is only one component of the random graph with more than* $\beta \log n$ *vertices. The size of this component* $\sim (1 - \rho(\lambda))n$ *where* $\rho(\lambda)$ *is the extinction probability for the Poisson(λ) branching process.*

Remark. This time our constant is the end result of several choices and is not so good. Corollary 5.11 of Bollobás (2001) shows that when $\lambda > 1$ the largest nongiant component is asymptotically $(1/\alpha)\log n$ where $\alpha = \lambda - 1 - \log(\lambda) > 0$. We will prove that result in Theorem 2.6.4.

Proof. There are four steps.

Step 1. There is a constant $\gamma = 2/\theta_\delta$ so that if $W_0 \geq \gamma \log n$ then the probability W_t ever hits 0 is $\leq n^{-2}$.

Step 2. Using $|A_t| \approx S_t$ the random walk from the previous section, we show there is a constant β so that $P(0 < |A(\beta \log n)| < \gamma \log n) = o(n^{-1})$.

Step 3. Using $W_t \leq |A_t| \leq S_t$, we show that with probability $\geq 1 - \exp(-\eta n^{2/3})$ we have $\epsilon n^{2/3} \leq |A(n^{2/3})| \leq 2\lambda n^{2/3}$. Combined with the first two steps, this shows that with probability $\to 1$, all clusters reaching size $\beta \log n$ will intersect producing a giant component.

Step 4. We show that the number of sites with clusters of size $\leq \beta \log n$ is asymptotically ρn.

Step 1. The increments of W have the distribution $-1 + \text{Binomial}((1 - \delta)n, \lambda/n)$. By (2.3.3) the moment generating function of an increment

$$\phi_\delta(\theta) \leq \exp(-\theta + \lambda(1 - \delta)(e^\theta - 1)) \equiv \psi_\delta(\theta) \tag{2.3.6}$$

Choose $\delta > 0$ so that $\lambda(1 - \delta) > 1$. $\psi_\delta'(0) = -1 + \lambda(1 - \delta) > 0$ so $\psi_\delta(-\theta) < 1$ when $\theta > 0$ is small. Since $\psi_\delta(-\theta)$ is convex and tends to ∞ as $\theta \to \infty$ there is a unique positive solution of $\psi_\delta(-\theta_\delta) = 1$. $\phi_\delta(-\theta_\delta) \leq 1$ so $M_t = \exp(-\theta_\delta W_t)$ is a nonnegative supermartingale. Let $T_0 = \inf\{t : W_t = 0\}$. Stopping at time $T_0 \wedge t$ we have

$$e^{-\theta_\delta} \geq E_1(\exp(-\theta_\delta W(T_0)); T_0 \leq t) = P_1(T_0 \leq t)$$

Letting $t \to \infty$ we have

$$e^{-\theta_\delta} \geq P_1(T_0 < \infty) \tag{2.3.7}$$

To compare with the previous approach using branching processes, note that $\rho = e^{-\theta_0} < 1$ has $\theta_0 + \lambda(e^{-\theta_0} - 1) = 0$ (since $\psi_0(-\theta_0) = 1$) so rearranging gives

$$\rho = \exp(\lambda(\rho - 1))$$

which is the fixed point equation for the branching process.

From the proof of (2.3.7) it is immediate that if P_m denotes the probability when $W_0 = m$

$$P_m(T_0 < \infty) \leq e^{-\theta_\delta m} \tag{2.3.8}$$

We want to make sure that all of the components $\mathcal{C}_x$, $1 \leq x \leq n$ behave as we expect, so we take $m_\delta = (2/\theta_\delta) \log n$ to make the right-hand side n^{-2}.

Large Deviations Bound. To control the behavior of W_t and S_t with good error bounds we use the following:

Lemma 2.3.3. *Let* $Z = X_1 + \cdots + X_t$ *where the* X_i *are independent* $Binomial((1 - \delta)n, \lambda/n)$. *Let* $\mu = EZ = t\lambda(1 - \delta)$. *Let* $\gamma(x) = x \log x - x + 1$ *which is* > 0 *when* $x \neq 1$. *If* $x < 1 < y$ *then* $P(Z \leq x\mu) \leq e^{-\gamma(x)\mu}$ *and* $P(Z \geq y\mu) \leq e^{-\gamma(y)\mu}$.

Proof. $E \exp(\theta Z) \leq \exp(\mu(e^\theta - 1))$ by (2.3.6). If $\theta > 0$ Markov's inequality implies

$$P(Z \geq y\mu) \leq \exp((-\theta y + e^\theta - 1)\mu)$$

Since $y > 1$, $-\theta y + e^\theta - 1 < 0$ for small $\theta > 0$. Differentiating we see that the bound is optimized by taking $\theta = \log y$. If $\theta < 0$ Markov's inequality implies

$$P(Z \leq x\mu) \leq \exp((-\theta x + e^\theta - 1)\mu)$$

Since $x < 1$, $-\theta x + e^\theta - 1 < 0$ for small $\theta < 0$. Differentiating we see that the bound is optimized by taking $\theta = \log x$. ∎

Step 2. Let $\epsilon = (\lambda - 1)/2$. Applying Lemma 2.3.3 to $S_t - S_0 + t$ with $x = (1+\epsilon)/\lambda$ we see that there is an $\eta_1 > 0$ so that

$$P(S_t - S_0 \leq \epsilon t) \leq \exp(-\eta_1 t) \tag{2.3.9}$$

Taking $y = 2$ in Lemma 2.3.3 we have an $\eta_2 > 0$ so that

$$P(S_t - S_0 + t \geq 2\lambda t) \leq \exp(-\eta_2 t) \tag{2.3.10}$$

Recall $S_0 = 1$. When $S_t + t \leq 2\lambda t$, we have $|U_s| \geq n - 2\lambda t$ for all $s \leq t$, so the number of births lost in A_s for $s \leq t \wedge \tau$

$$\leq \sum_{s=0}^{t-1} \sum_{y=|U_s|+1}^{n} \zeta_y^s \leq \text{Binomial}(2\lambda t^2, \lambda/n) \tag{2.3.11}$$

From this it follows that

$$P(|A_t| > 0, S_t \neq |A_t|) \leq 2\lambda^2 t^2/n \tag{2.3.12}$$

$$P(|A_t| > 0, S_t - |A_t| \geq 2) \leq \binom{2\lambda t^2}{2} \left(\frac{\lambda}{n}\right)^2 \tag{2.3.13}$$

We are going to use the results in the previous paragraph at time $r = \beta \log n$ where β is chosen so that $\beta\epsilon \geq 3/\theta_\delta$, $\beta\eta_1 \geq 2$ and $\beta\eta_2 \geq 2$. Combining (2.3.9), (2.3.10), and (2.3.13) we have

$$P(0 < |A_r| \leq (3/\theta_\delta)\log n - 2) \leq C\frac{(\log n)^2}{n^2}$$

If n is large $(3/\theta_\delta)\log n - 2 \geq (2/\theta_\delta)\log n = m_\delta$, so (2.3.8) implies that if $|A_r| > 0$ it is unlikely that the lower bounding random walk will ever hit 0.

Step 3. Let $\epsilon_\delta = (\lambda(1-\delta) - 1)/2$. Using Lemma 2.3.3 twice we have

$$P_1\left(W\left(n^{2/3}\right) - W(0) \leq \epsilon_\delta n^{2/3}\right) \leq \exp\left(-\eta_3 n^{2/3}\right)$$
$$P_1\left(W\left(n^{2/3}\right) - W(0) + n^{2/3} > 2\lambda n^{2/3}\right) < \exp\left(-\eta_4 n^{2/3}\right) \tag{2.3.14}$$

Here we take $W(0) = |A_r| \le 2\lambda\beta\log n$. Since $W_t + t$ is nondecreasing this shows that with probability $1 - O(n^{-2})$, $W_s + s \le \delta n$ for all $s \le n^{2/3}$, and the coupling between W_s and $|A(s+r)|$ remains valid for $0 \le s \le n^{2/3}$.

The first bound in (2.3.14) implies that if a cluster reaches size $r = \beta\log n$ then the set of active sites at time $r + n^{2/3}$ is $\ge \epsilon_\delta n^{2/3}$ with high probability. Thus if we have two such clusters of size $\ge \beta\log n$ then either (a) they will intersect by time $r + n^{2/3}$ or (b) at time $r + n^{2/3}$ they have disjoint sets I and J of active sites of size $\ge \epsilon_\delta n^{2/3}$. The probability of no edge connecting I and J is

$$\le \left(1 - \frac{\lambda}{n}\right)^{\epsilon_\delta^2 n^{4/3}} \le \exp(-\eta_5 n^{1/3})$$

where $\eta_5 = \lambda\epsilon_\delta^2$. This proves the first assertion in Theorem 2.3.2.

Step 4. To prove the second assertion it suffices to show that

$$|\{x : |\mathcal{C}_x| \le \beta\log n\}|/n \to \rho(\lambda) \tag{2.3.15}$$

The first step is to show $P(|\mathcal{C}_x| \le \beta\log n) \to \rho(\lambda)$. Let $T_0 = \inf\{t : S_t = 0\}$. Because $S_t \ge |A_t|$, the probability in question is

$$\ge P(T_0 \le \beta\log n) - o(1) \to \rho(\lambda)$$

For the other direction we note that (2.3.13) shows that $P(T_0 > \beta\log n, S_t \ne |A_t|) \to 0$.

To complete the proof of (2.3.15) we will show that the random variables $Y_x = 1$ if $|\mathcal{C}_x| \le \beta\log n$ and 0 otherwise are asymptotically uncorrelated. We isolate the reasoning as

Lemma 2.3.4. *Let F be an event that involves exposing J vertices starting at 1, and let G be an event that involves exposing K vertices starting at 2. Then*

$$|P(F \cap G) - P(F)P(G)| \le JK \cdot \frac{2\lambda}{n}$$

Proof. Let R_t, U_t, and A_t be the process of exposing the cluster of 1. Introduce independent copies of the basic indicator random variables $\eta'_{x,y}$. Let $R'_0 = \emptyset$, $A'_0 = \{2\}$ and $U'_0 = \{1, 2, \ldots, n\} - \{2\}$. If $A'_t \ne \emptyset$, pick $i'_t = \min A'_t$. If $i'_t \notin R_J$ let

$$\begin{aligned}
R'_{t+1} &= R'_t \cup \{i'_t\} \\
A'_{t+1} &= A'_t - \{i'_t\} \cup \{y \in U'_t : \eta'_{i'_t,y} = 1\} \\
U'_{t+1} &= U'_t - \{y \in U'_t : \eta'_{i'_t,y} = 1\}
\end{aligned} \tag{2.3.16}$$

However, if $i_t' \in R_J$, an event we call a collision, we use $\eta_{i_t,y}$ instead of $\eta_{i_t,y}'$. In words if while growing cluster 2 we choose a site that was used in the growth of cluster 1, we use the original random variables $\eta_{x,y}$. Otherwise we use independent random variables.

It should be clear from the construction that

$$P(F \cap G) - P(F)P(G) \le P(R_J \cap R_K' \ne \emptyset) \le JK \cdot \frac{\lambda}{n}$$

and this completes the proof. ∎

Using Lemma 2.3.4 with $J = K = \beta \log n$, the probability of a collision is at most $\lambda(\beta \log n)^2 / n$. Using our bound on the covariance

$$\mathrm{var}\left(\sum_{x=1}^{n} Y_x\right) \le n + \binom{n}{2}\frac{\lambda(\beta \log n)^2}{n} \le Cn \log^2 n$$

so it follows from Chebyshev's inequality that

$$P\left(\sum_{x=1}^{n}(Y_x - EY_x) \ge n^{2/3}\right) \le \frac{Cn \log^2 n}{n^{4/3}} \to 0$$

This proves (2.3.15) and completes the proof of Theorem 2.3.2. ∎

2.4 Diameter of the Giant Component

Having proved the existence of the giant component, we can use the branching process results from Section 2.1 to study the typical distance between two points on the giant component.

Theorem 2.4.1. *Suppose $\lambda > 1$ and pick two points x and y at random from the giant component. Then $d(x, y)/\log n \to 1/\log \lambda$ in probability.*

The answer in Theorem 2.4.1 is intuitive. The branching process approximation grows at rate λ^t, so the average distance is given by solving $\lambda^t = n$, that is, $t = (\log n)/\log \lambda$.

Proof. We begin with a small detail. The growth of the cluster containing x in the graph on n can be approximated by a branching process Z_t^n. Unfortunately, the offspring distribution Binomial$(n, \lambda/n)$ depends on n, so if we are concerned with the behavior of $Z^n(t_n)$ with $t_n \to \infty$ we have a triangular array, not a single sequence. To deal with this, we use (6.4) in Chapter 2 of Durrett (2004) to conclude that the total variation distance between Binomial$(n, \lambda/n)$ and Poisson(λ)

is $\leq 2\lambda^2/n$. It follows from the definition of the total variation distance that $\xi =$ Binomial$(n, \lambda/n)$ and $\xi' =$ Poisson(λ) can be constructed on the same space in such a way that $P(\xi \neq \xi') \leq 2\lambda^2/n$. In what follows we will run the branching process until the total number of birth events is $o(n)$ so with a probability that $\to 1$ there are no differences between the Binomial and Poisson versions. Thus we can use the growth results for the Poisson process and make conclusions about the Binomial one.

The size of $Z_t \geq |I_t|$ at time $(1 - \epsilon)\log n/\log \lambda$ is $\sim n^{1-\epsilon} W$, so most of the points are at least distance $\geq (1 - \epsilon)\log n/\log \lambda$. As Theorem 2.3.2 shows, membership in the giant component is asymptotically equivalent to the cluster size being larger than $\beta \log n$. Let Y_∞ be the total progeny of the branching process. When $\lambda > 1$, $P(\beta \log n < Y_\infty < \infty) \to 0$ as $n \to \infty$, so points in the infinite cluster are with high probability associated with branching processes that don't die out. If δ is small then the size of one of these processes at time $(1 + \epsilon)\log n/(2\log \lambda)$ is $\sim n^{(1+\epsilon)/2}\delta$ with high probability. If we consider two growing clusters at this time then either (a) they have already intersected or (b) they will have $n^{1+\epsilon}\delta^2$ chances to intersect at the next time, and the probability they will fail is

$$\leq \left(1 - \frac{\lambda}{n}\right)^{n^{1+\epsilon}\delta^2} \leq \exp(-n^\epsilon \delta^2) \to 0$$

and the proof is complete.   ∎

Our next task is to show that the diameter $D = \max d(x, y)$ of the giant component has a different limiting behavior than the average distance between two randomly chosen points. A *dangling end* is path $v_0, v_1, \ldots v_{k+1}$ with the degree $d(v_0) = 1$ and $d(v_i) = 2$ for $1 \leq i \leq k$.

Theorem 2.4.2. *Let* $p_1 = e^{-\lambda}\lambda$ *and choose* $k(n)$ *so that* $c_n = np_1^{k(n)+1}$ *stays bounded away from* 0 *and* ∞. *When* n *is large the probability of a dangling end of length* $k(n)$ *is approximately* $1 - \exp(-c_n)$.

Here $k(n) = \log n/(\log p_1) + O(1)$. The statement is made complicated by the fact that increasing $k(n)$ by 1 decreases $np_1^{k(n)+1}$ by a factor p_1, so we cannot choose $k(n)$ to make c_n converge to $c \in (0, \infty)$.

Proof. Let $A_{(x,y)}$ be the event that there is dangling end $v_0 = x, v_1, \ldots v_{k+1} = y$

$$P(\cup_{(x,y)} A_{(x,y)}) \leq \sum_{(x,y)} P(A_{(x,y)}) \sim n(n - 1) \cdot p_1^{k+1} \cdot \frac{1}{n - 1} = c_n$$

To see this note that there are $n(n - 1)$ values for the ordered pair (x, y) with $y \neq x$. v_0 has degree 1, an event of probability p_1. As for the v_i with $i > 0$, we arrive there along the edge from v_{i-1} and we have $n - i$ uninspected outgoing edges so the probability of finding exactly one open is $\sim p_1$. When $i < k$ we don't care what vertex we connect to, but when $i = k$ we want v_k to connect to y, an event that by symmetry has probability $P_1/(n - 1)$.

By the inclusion–exclusion formula we can get a lower bound by subtracting the sum of $P(A_{(x,y)} \cap A_{(w,z)})$ for all $(x, y) \neq (w, z)$. It is easy to see that if $y \neq z$, $A_{(x,y)} \cap A_{(x,z)} = \emptyset$. In all the other cases the paths for these two events cannot intersect, except perhaps at $y = z$. This gives the second event less room to occur, so in all cases

$$P(A_{(x,y)} \cap A_{(w,z)}) \leq P(A_{(x,y)}) \cdot P(A_{(w,z)}) \qquad (2.4.1)$$

Combining the two estimates we have

$$\sum_{(x,y) \neq (w,z)} P(A_{(x,y)} \cap A_{(w,z)}) \leq \frac{1}{2}(n(n - 1))^2 P(A_{(x,y)})^2 \sim c_n^2/2$$

Before progressing to the third level of approximation, we need to improve (2.4.1) by observing that the occurrence of $A_{x,y}$ in effect removes $k + 1$ vertices from the graph so unless $x = w$ and $y \neq z$,

$$P(A_{(x,y)} \cap A_{(w,z)})/P(A_{(x,y)}) \cdot P(A_{(w,z)}) \to 1$$

as $n \to \infty$. This result generalizes easily to a fixed finite number of events. Using this with the mth approximation to the inclusion–exclusion formula we get an approximation

$$c_n - c_n^2/2! + \cdots + (-1)^{m+1} c_n^m/m!$$

which is an upper bound or a lower bound depending on whether the last term is $+$ or $-$. The infinite series is $1 - \exp(-c_n)$ and the desired result follows. ∎

Suppose we have two dangling ends, one from x to y and the other from w to z, each of length $a \log n$. Conditioning on their existence removes some vertices from the graph but does not significantly change the growth of clusters from y and z. Thus in $ER(n, \lambda/n)$, we have positive probability of y and z belonging to the giant component. When they do, they will have distance $\approx (\log n)/(\log \lambda)$, and x and w will have distance $2a \log n + (\log n)/(\log \lambda)$. Chung and Lu (2001) have obtained upper bounds on the diameter of the giant component, see Theorem 6 on their page 272.

Looking at the last paragraph, a mathematician might conclude that the diameter is what we should be studying, because it is harder to understand than the typical distance. While the diameter is the more traditional notion, there are two good reasons to study the distance between two randomly chosen points. The first is that the diameter is a number while the distribution of the typical distance contains substantially more information. In addition, the diameter is rather sensitive to small changes in the graph: adding a dangling end can substantially change the diameter, but will not affect the typical distance.

2.5 CLT for the Giant Component

Up to this point we have been content to study the growth of clusters while they are $o(n)$. In this section we will use an idea of Martin-Löf (1986) to follow the random walk approach all of the way to the end of the formation of the giant component and prove a central limit theorem for the size of the giant component.

To avoid the problem caused by the process dying out, it is convenient to modify the rules so that if $A_t = \emptyset$ we pick $i_t \in U_t$, and rewrite the recursion as

$$R_{t+1} = R_t \cup \{i_t\}$$
$$A_{t+1} = A_t \cup \{y \in U_t : \eta_{i_t,y} = 1\}$$
$$U_{t+1} = U_t - (\{i_t\} \cup \{y \in U_t : \eta_{i_t,y} = 1\})$$

In words, when one cluster is finished we pick a new vertex and start exposing its cluster.

When $A_t = \emptyset$ we subtract $1 + \text{Binomial}(|U_t|, \lambda/n)$ points from U_t versus $\text{Binomial}(|U_t|, \lambda/n)$ points when $A_t \neq \emptyset$. However, we will experience only a geometrically distributed number of failures before finding the giant component, so this difference can be ignored. Let $\mathcal{F}_t$ be the σ-field generated by the process up to time t. Let $u_t^n = |U_t|$.

Lemma 2.5.1. $u_{[ns]}^n/n$ *converges in distribution to* u_s *the solution of*

$$\frac{du_s}{ds} = -\lambda u_s \qquad u_0 = 1$$

and hence $u_s = \exp(-\lambda s)$.

Proof. Let $\Delta u_t^n = u_{t+1}^n - u_t^n$. If $A_t \neq \emptyset$ then

$$E(\Delta u_t^n | \mathcal{F}_t) = -u_t^n \frac{\lambda}{n}$$

$$\text{var}(\Delta u_t^n | \mathcal{F}_t) = u_t^n \frac{\lambda}{n}\left(1 - \frac{\lambda}{n}\right)$$

If we let $t = [ns]$ for $0 \leq s \leq 1$ and divide by n then

$$E\left(\frac{\Delta u_{[ns]}^n}{n} \,\middle|\, \mathcal{F}_{[ns]}\right) = -\frac{u_{[ns]}^n}{n} \cdot \lambda \cdot \frac{1}{n}$$

$$\text{var}\left(\frac{\Delta u_{[ns]}^n}{n} \,\middle|\, \mathcal{F}_{[ns]}\right) = \frac{u_{[ns]}^n}{n} \cdot \lambda \left(1 - \frac{\lambda}{n}\right) \cdot \frac{1}{n^2} \qquad (2.5.1)$$

Dividing each right-hand side by $1/n$, the time increment in the rescaled process, we see that $\Delta u^n_{[ns]}$ has

$$\text{infinitesimal mean} = -\frac{u^n_{[ns]}}{n}\lambda$$

$$\text{infinitesimal variance} = \frac{u^n_{[ns]}}{n}\lambda\left(1-\frac{\lambda}{n}\right)\cdot\frac{1}{n}$$

Letting $n \to \infty$, the infinitesimal variance $\to 0$, so the result follows from (7.1) in Chapter 8 of Durrett (1996). ∎

The last proof is simple and intuitive, but may be too sophisticated for some reader's tastes, so we now give

Alternative proof. The calculations above show that

$$M^n_t = \left(1-\frac{\lambda}{n}\right)^{-t} u^n_t/n$$

is a martingale with

$$E(M^n_t - M^n_0)^2 = \sum_{s=0}^{t-1} E\left(M^n_{s+1} - M^n_s\right)^2$$
$$\leq \sum_{s=0}^{t-1} \left(1-\frac{\lambda}{n}\right)^{-2s+1} \lambda/n^2 \to 0$$

so by Kolmogorov's maximal inequality

$$E\left(\max_{0\leq s\leq n}(M^n_s - M^n_0)^2\right) \to 0$$

Since $M^n_0 = 1$, this says that when n is large $M^n_s \approx 1$ uniformly in s, so $u^n_{[ns]}/n \approx (1-\lambda/n)^{[ns]} \to e^{-\lambda s}$. ∎

To determine the size of the giant component, we note that when $u^n_t + r^n_t = n$, $A_t = \emptyset$. This may occur several times for small t while we are searching for the giant component, but the solution we are looking for is the first occurrence after an excursion of $O(n)$. To locate roughly the time at which this occurs, we note that scaling $r^n_t = |R_t| \equiv t$ as we did u^n_t, $r^n_{[ns]}/n \to s$. [Here and in what follows we will use t for the original integer time scale and $s \in [0, 1]$ for rescaled time.] After scaling

$$u^n_t + r^n_t = n \quad\Rightarrow\quad e^{-\lambda s} + s = 1$$

Solving we have $1 - s = \exp(\lambda((1 - s) - 1))$, which is the fixed point equation for the extinction probability, $1 - s$. As the graph below shows $e^{-\lambda s} + s > 1$ for $s > 1 - \rho$, so we are interested only in $u^n_{[ns]}/n$ for $0 \leq s \leq 1 - \rho + \epsilon$. In this part of the process we first generate a geometrically distributed number of small clusters and then expose the giant component.

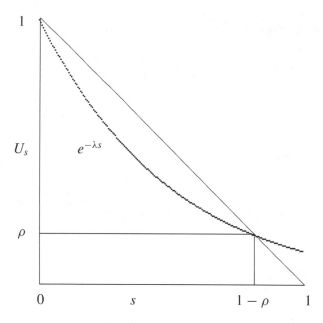

Example with $\lambda = 2$.

Consider now $y^n_{[ns]} = (u^n_{[ns]} - n\exp(-\lambda s))/\sqrt{n}$ for $0 \le s \le 1 - \rho$.

Lemma 2.5.2. *As $n \to \infty$, $y^n_{[ns]}$ converges in distribution to a normal with mean 0 and variance $e^{-\lambda s} - e^{-2\lambda s}$.*

Proof. If $A_{[ns]} \ne \emptyset$ then using the formulas in (2.5.1)

$$E(\Delta y^n_{[ns]} | \mathcal{F}_{[ns]}) = \frac{1}{\sqrt{n}}\left(-u^n_{[ns]} \cdot \frac{\lambda}{n} - n\exp(-\lambda s)(\exp(-\lambda/n) - 1)\right)$$

$$\sim -\frac{\lambda}{n}\left(\frac{u^n_{[ns]} - n\exp(-\lambda s)}{\sqrt{n}}\right) = -\lambda y^n_{[ns]} \cdot \frac{1}{n}$$

$$\mathrm{var}\,(\Delta y^n_{[ns]} | \mathcal{F}_{[ns]}) = \mathrm{var}\left(\frac{\Delta u^n_{[ns]}}{\sqrt{n}}\,\bigg|\,\mathcal{F}_{[ns]}\right) = \frac{1}{n} \cdot u^n_{[ns]} \cdot \frac{\lambda}{n}\left(1 - \frac{\lambda}{n}\right) \sim \lambda e^{-\lambda s} \cdot \frac{1}{n}$$

Using (7.1) in Chapter 8 of Durrett (1996) again, we see that $y^n_{[ns]}$ converges in distribution to the solution of the following stochastic differential equation.

$$dy_s = -\lambda y_s\, ds + \sqrt{\lambda e^{-\lambda s}}\, dB_s \quad y_0 = 0$$

The solution to this equation is

$$y_s = \int_0^s e^{-\lambda(s-r)}\sqrt{\lambda e^{-\lambda r}}\, dB_r \tag{2.5.2}$$

To check this, recall that if one continuously invests an amount g_s in an exponentially decaying stock market then your net wealth x_s satisfies

$$\frac{dx_s}{ds} = -\lambda x_s + g_s$$

Since computation of interest is linear, each amount decays exponentially from its date of investment, and this differential equation has solution

$$x_s = \int_0^s e^{-\lambda(s-r)} g_r \, dr$$

Readers who want a more rigorous proof can use stochastic calculus to check this.

Since the integrand in (2.5.2) is deterministic, y_s has a normal distribution with mean 0 and variance

$$\int_0^s \exp(-2\lambda(s-r))\lambda e^{-\lambda r} \, dr = e^{2\lambda s} \int_0^s \lambda e^{\lambda r} \, dr = e^{-\lambda s} - e^{-2\lambda s}$$

which proves the result. ∎

Remark. Again if one wants to avoid stochastic calculus, the theorem can be proved by applying the martingale central limit theorem to

$$M_t^n - M_0^n = \left(1 - \frac{\lambda}{n}\right)^{-t} u_t^n/n - 1$$

The key observation is that

$$\sum_{r=0}^{[ns]-1} E((M_{r+1}^n - M_r^n)^2 | \mathcal{F}_r) \to \int_0^s \lambda e^{\lambda u} \, du = e^{\lambda s} - 1$$

that is, the variance process has a deterministic limit. See, for example, Section 7.7 in Durrett (2004). Multiplying the martingale by $e^{-\lambda s}$, multiplies the variance by $e^{-2\lambda s}$ and we arrive at the same limit as before.

We have analyzed the fluctuations of $u_{[ns]}^n$. To determine the fluctuations of the point where $u_t^n + t = n$, we can now prove the result as we do the central limit theorem of renewal theory. To briefly recall that approach, let $\xi_1, \xi_2, \ldots$ be i.i.d. positive random variables with $E\xi_i = \mu$ and $\text{var}(\xi_i) = \sigma^2 \in (0, \infty)$. Let $T_n = \xi_1 + \cdots + \xi_n$ and $N(t) = \inf\{n : T_n > t\}$. The central limit theorem implies

$$T_n \approx n\mu + \sigma \sqrt{n}\chi$$

where χ is a standard normal. Setting $n = t/\mu$

$$T_{t/\mu} \approx t + \sigma \sqrt{\frac{t}{\mu}}\chi$$

If $\chi > 0$ then $N(t) < t/\mu$. The law of large numbers implies $T_n - T_m \approx (n - m)\mu$ when $n - m$ is large so we will have

$$\frac{t}{\mu} - N(t) \approx \frac{\sigma}{\mu}\sqrt{\frac{t}{\mu}}\chi$$

The same reasoning applies in the current situation. Taking $s = 1 - \rho$ in Lemma 2.5.2 and letting $\mathcal{Z}$ denote a normal with mean 0 and variance $e^{-\lambda(1-\rho)} - e^{-2\lambda(1-\rho)}$ we have

$$u^n_{[n(1-\rho)]} \approx n\exp(-\lambda(1-\rho)) + \sqrt{n}\mathcal{Z} \qquad (2.5.3)$$

$A_{[ns]} = \emptyset$ when $u^n_{[ns]} = n - [ns]$. To find out when this occurs, we suppose equality holds at $s_0 = (1 - \rho) + \mathcal{Y}/\sqrt{n}$. Using (2.5.3) and noting $s_0 \to (1 - \rho)$ as $n \to \infty$

$$n\exp(-\lambda\{(1 - \rho) + \mathcal{Y}/\sqrt{n}\}) + \sqrt{n}\mathcal{Z} = u^n_{[ns_0]} = n - [ns_0]$$

or rearranging

$$\exp(-\lambda s_0) - 1 + s_0 = -\mathcal{Z}/\sqrt{n}$$

Let $h(t) = e^{-\lambda t} - 1 + t$ which is $= 0$ at $t = 1 - \rho$. $h'(t) = -\lambda e^{-\lambda t} + 1$, so we can write the above as

$$h'(1 - \rho)\mathcal{Y}/\sqrt{n} \approx -\mathcal{Z}/\sqrt{n}$$

or $\mathcal{Y} \approx \mathcal{Z}/h'(1 - \rho)$. $h'(1 - \rho) = 1 - \lambda\rho$. Putting the pieces together.

Theorem 2.5.3. *Suppose $\lambda > 1$. The size of the largest component $\mathcal{C}^{(1)}$ satisfies*

$$\frac{|\mathcal{C}^{(1)}| - n(1 - \rho)}{\sqrt{n}} \Rightarrow \chi$$

where $\Rightarrow$ means convergence in distribution and χ has a normal distribution with mean 0 and variance $(\rho - \rho^2)/(1 - \lambda\rho)^2$.

For other approaches to this result see Pittel (1990) and Barraez, Boucheron, and Fernandez de la Vega (2000). To compare variances note that Pittel's $c = \lambda$ and $T = \rho/\lambda$.

2.6 Combinatorial Approach

Combinatorial methods give more refined results about the Erdős–Rényi model, but they also work more easily in the presence of qualitative results like Theorems 2.3.1 and 2.3.2 and the next lemma. First, we need some graph theoretic preliminaries. It is easy to see that a tree with k vertices has $k - 1$ edges. We call a graph with

k vertices and k edges, a *unicyclic graph*, since it will have exactly one cycle, that is, a path of adjacent vertices $x_0, x_1, \ldots, x_k = x_0$, and $x_j \neq x_0$ for $1 \leq j < k$. We call a graph with k vertices and $k + \ell$ edges with $\ell \geq 1$ a *complex component* with complexity ℓ.

Our first result, says that complex components are rare, except possibly near the critical value $\lambda = 1$.

Lemma 2.6.1. *Suppose $\lambda \neq 1$, let $A < \infty$ be a constant, and consider only components with $\leq A \log n$ vertices. The expected number of unicyclic components is $\leq \lambda (A \log n)^2$. The probability of having at least one complex component is $\leq (A \log n)^4 \lambda^2 / n$.*

Proof. We look at the growth of the cluster from the random walk viewpoint. An increase in complexity is caused by a "self-intersection," that is, a connection from i_t to some point in A_t. Our assumption implies that there are $\leq A \log n$ times and $|A_t| \leq A \log n$ for all t so the number of self-intersections is $\leq$ Binomial$((A \log n)^2, \lambda / n)$. The result for unicyclic components follows by recalling there are n possible starting points and taking expected value. The result for complex components follows by computing the probability of two or more self-intersections as we did in (2.3.13). ∎

Having ruled out complex components, we can, for $\lambda \neq 1$, restrict our attention to tree and unicyclic components. Cayley showed in 1889 that there are k^{k-2} trees with k-labeled vertices. When $p = \lambda / n$ the expected number of trees of size k present is

$$\binom{n}{k} k^{k-2} \left(\frac{\lambda}{n}\right)^{k-1} \left(1 - \frac{\lambda}{n}\right)^{k(n-k)+\binom{k}{2}-(k-1)} \tag{2.6.1}$$

since each of the $k - 1$ edges in the tree needs to be present and there can be no edges connecting its k vertices to its complement or any other edges connecting the k vertices. For fixed k, we can drop $-k^2 + \binom{k}{2} - k + 1$ from the exponent of the last term and the above is asymptotic to

$$n \frac{k^{k-2}}{k!} \lambda^{k-1} e^{-\lambda k} \equiv n q_k \tag{2.6.2}$$

Recalling that in the subcritical regime cluster sizes have the same distribution as the total progeny in a Poisson(λ) branching process. we get the following corollary, which is "well-known," but not easy to prove directly from the definition of the branching process.

Corollary 2.6.2. *The probability distribution of the total progeny τ of a* Poisson(λ) *branching process with $\lambda < 1$ is given by*

$$P(\tau = k) = kq_k = \frac{1}{\lambda}\frac{k^{k-1}}{k!}(\lambda e^{-\lambda})^k \tag{2.6.3}$$

There is an extra factor of k due to the fact that a tree of size k is C_x for k values of x.

This distribution was first discovered by Borel (1942). He showed that when $\lambda < 1$ it gave the distribution of the total number of customers served in the first busy period of a queue with Poisson rate λ arrivals and service times always equal to 1. It is called the Borel-Tanner distribution, since Tanner (1953) extended to the case that the initial queue size is arbitrary (see also Tanner (1961)). Of course, this becomes a branching process if we think of the customers that arrive during a person's service time as their children.

Duality. Suppose $\lambda > 1$ and let ρ be the extinction probability. The title of this subsection refers to the fact that there is a close relationship between Erdős–Rényi random graphs with mean degrees $\lambda > 1$ and $\lambda\rho < 1$. Using the fixed point equation $\rho = e^{\lambda(\rho-1)}$ implies

$$\lambda\rho e^{-\lambda\rho} = \lambda e^{\lambda(\rho-1)}e^{-\lambda\rho} = \lambda e^{-\lambda} \tag{2.6.4}$$

Let $ER(n, p)$ denote an Erdős–Rényi graph with n vertices and edges present with probability p. Let $m = n\rho$ and consider $ER(m, \lambda\rho/m)$, an Erdős–Rényi graph with number of vertices asymptotically equal to the number of vertices in nongiant components of $ER(n, \lambda/n)$. Changing variables in (2.6.2) we see that

$$\frac{m}{\lambda\rho}\frac{k^{k-2}}{k!}(\lambda\rho e^{-\lambda\rho})^k = \frac{n}{\lambda}\frac{k^{k-2}}{k!}(\lambda e^{-\lambda})^k$$

In words, the expected number of trees of size k is the same in $ER(m, \lambda\rho/m)$ and $ER(n, \lambda/n)$. Changing variables in the same way in (2.6.3)

$$\frac{1}{\lambda\rho}\frac{k^{k-1}}{k!}(\lambda\rho e^{-\lambda\rho})^k = \frac{1}{\lambda}\frac{k^{k-1}}{k!}(\lambda e^{-\lambda})^k \cdot \frac{1}{\rho}$$

In words, the total progeny of a Poisson(λ) branching process conditioned on extinction is the same as that of a Poisson($\lambda\rho$) branching process, which is Theorem 2.1.8

Equation (2.6.2) is a result about the expected number of trees. The next result is a law of large numbers, which says that the actual number is close to the expected value.

Theorem 2.6.3. *Let T_k^n be the number of tree components of size k in the Erdős–Rényi graph with n vertices. As $n \to \infty$, $T_k^n/n \to q_k$ in probability, where q_k is defined in (2.6.2)*

Proof. This proof comes from Bollobás (2001, pp. 106–107). The expected number of ordered pairs of tree components of size k (with the second tree different from the first) is

$$\binom{n}{k} k^{k-2} \left(\frac{\lambda}{n}\right)^{k-1} \left(1 - \frac{\lambda}{n}\right)^{k(n-k)+\binom{k}{2}-k+1}$$

$$\times \binom{n-k}{k} k^{k-2} \left(\frac{\lambda}{n}\right)^{k-1} \left(1 - \frac{\lambda}{n}\right)^{k(n-2k)+\binom{k}{2}-k+1} \tag{2.6.5}$$

The second formula differs from the first only in two places: first we have only $n - k$ vertices to choose from, and the first term already takes into account the fact that there are no connections from the first tree to the second. Since $\binom{n-k}{k} \leq \binom{n}{k}$ the above is

$$\leq \left(ET_k^n\right)^2 \left(1 - \frac{\lambda}{n}\right)^{-k^2} \leq \left(ET_k^n\right)^2 e^{\lambda k^2/n}$$

From this we get

$$\operatorname{var}\left(T_k^n\right) = E\left(T_k^n\left(T_k^n - 1\right)\right) + ET_k^n - \left(ET_k^n\right)^2 \leq ET_k^n + \left(ET_k^n\right)^2 \left(e^{\lambda k^2/n} - 1\right)$$

Using Chebyshev's inequality

$$P\left(|T_k^n - ET_k^n| \geq n^{2/3}\right) \leq \frac{ET_k^n + \left(ET_k^n\right)^2 \left(e^{\lambda k^2/n} - 1\right)}{n^{4/3}} \to 0$$

since $ET_k^n \sim nq_k$ and $e^{\lambda k^2/n} - 1 \sim \lambda k^2/n$. This gives the desired result. Note that we could replace $n^{2/3}$ in the last display by $\omega(n)n^{1/2}$ where $\omega(n) \to \infty$ as $n \to \infty$. ∎

   The results above allow us to verify the remark we made about the largest nongiant component for $\lambda > 1$.

Theorem 2.6.4. *Suppose $\lambda > 1$ and let $\mathcal{C}^{(2)}$ be the second largest component. If $\alpha = \lambda - 1 - \log \lambda$ and $a > 1/\alpha$ then as $n \to \infty$*

$$P\left(|\mathcal{C}^{(2)}| \geq a \log n\right) \to 0$$

Proof. For simplicity we will do our calculations for the limit (2.6.2) rather than for the exact formula (2.6.1). Stirling's formula tells us that

$$k! \sim k^{k+1/2} e^{-k} \sqrt{2\pi} \quad \text{as } k \to \infty$$

so we have (Lemma 2.7.1 will show this is valid for $k = o(n^{1/2})$)

$$q_k = \frac{1}{\lambda} \cdot \frac{k^{k-2}}{k!} (\lambda e^{-\lambda})^k \sim \frac{1}{\lambda \sqrt{2\pi}} k^{-5/2} (\lambda e^{1-\lambda})^k$$

Now $g(\lambda) \equiv \lambda e^{1-\lambda} = 1$ when $\lambda = 1$ and $g'(\lambda) = (1 - \lambda)e^{1-\lambda}$. Thus $g(\lambda)$ is increasing for $\lambda < 1$, decreasing for $\lambda > 1$, and has $g(\lambda) < 1$ when $\lambda \neq 1$. Summing and using the fact that $k^{-5/2}$ is decreasing and $\lambda e^{1-\lambda} < 1$

$$Q_K = \sum_{k=K}^{\infty} q_k \sim \frac{1}{\lambda\sqrt{2\pi}} K^{-5/2} \frac{(\lambda e^{1-\lambda})^K}{1 - \lambda e^{1-\lambda}}$$

Taking $K = a \log n$

$$(\lambda e^{1-\lambda})^{a \log n} = \exp((\log \lambda - \lambda + 1)a \log n) = n^{-(1+\epsilon)}$$

when $a = (1 + \epsilon)/\alpha$, which proves the desired result.   ■

Unicyclic Components. Let $\mathcal{U}_n(\lambda)$ be the number of unicyclic components in $ER(n, \lambda/n)$. Rényi showed in 1959, see Bollobás (2001, p. 119) for a proof, that the number of unicyclic graphs with k vertices is

$$v_k = \frac{(k-1)!}{2} \sum_{j=0}^{k-3} \frac{k^j}{j!}$$

and for large k

$$v_k \sim \left(\frac{\pi}{8}\right)^{1/2} k^{k-1/2} \tag{2.6.6}$$

Computing with this formula as we did with Cayley's formula for the number of trees, the expected number of unicyclic components of size k is

$$E\mathcal{U}_n^k(\lambda) = \binom{n}{k} v_k \left(\frac{\lambda}{n}\right)^k \left(1 - \frac{\lambda}{n}\right)^{k(n-k)+\binom{k}{2}-k} \tag{2.6.7}$$

As $n \to \infty$ the above converges to

$$E\mathcal{U}^k(\lambda) = \frac{v_k}{k!}(\lambda e^{-\lambda})^k \tag{2.6.8}$$

Using the last result with (2.6.6) and Stirling's formula

$$E\mathcal{U}^k(\lambda) \sim \frac{1}{4} k^{-1}(\lambda e^{1-\lambda})^k$$

This is summable for $\lambda \neq 1$. Thus, in contrast to the $O(\log^2 n)$ upper bound in Lemma 2.6.1, the expected number of unicyclic components, $E\mathcal{U}(\lambda)$ converges to a finite limit.

When $\lambda < 1$ we can get an explicit formula for $E\mathcal{U}(\lambda)$ from a different approach. Consider an evolving random graph in which at times of a rate $n/2$ Poisson process edges arrive and are connected to two randomly chosen sites (if a connection between the sites does not exist already). By standard results on thinning a Poisson process the rate at which edges appear between a fixed pair x, y with $x \neq y$ is

$(n/2)/\binom{n}{2} = 1/(n-1)$. Therefore the probability of an edge is vacant at time t is $\exp(-t/(n-1)) \sim 1 - (t/n)$.

Unicyclic components are formed when we pick two points in the same tree component and they are not already connected by an edge. Call a collision picking two points in the same tree component. If $f_k(t)$ is the fraction of vertices that belong to clusters of size k then the rate at which collisions occur at time t is

$$\frac{n}{2} \sum_k \frac{k-1}{n} f_k(t) \qquad (2.6.9)$$

Now $f_k(t) = P(|\mathcal{C}_1| = k)$, so $\sum_k k f_k(t) = E|\mathcal{C}_1| = 1/(1-t)$. Integrating we have for $\lambda < 1$ that the expected number of collisions satisfies

$$E|\kappa_n(\lambda)| = \frac{1}{2} \int_0^\lambda \frac{1}{1-t} - 1 \, dt = \frac{1}{2}(-\log(1-\lambda) - \lambda)$$

The number of edges for which there have been two arrivals has mean

$$\left(\frac{t}{n}\right)^2 \binom{n}{2}$$

Subtracting this from the expected number of collisions

$$E|\kappa_n(\lambda)| \to \frac{1}{2}(-\log(1-\lambda) - \lambda - \lambda^2)$$

From a result in Berestycki and Durrett (2006) we get

Theorem 2.6.5. *If $\lambda < 1$ then as $n \to \infty$, $\kappa_n(\lambda)$ converges to a Poisson distribution with mean $(-\log(1-\lambda) - \lambda)/2$ and $\mathcal{U}_n(\lambda)$ converges to a Poisson distribution with mean $h(\lambda) = (-\log(1-\lambda) - \lambda - \lambda^2)/2$*

Proof. Let $N_n(t)$ be the number of collisions up to time t. It follows from (2.6.9) that

$$N_n(t) - \frac{1}{2} \sum_k (k-1) f_k(t) \quad \text{is a martingale.}$$

In the limit as $n \to \infty$, the compensator

$$\frac{1}{2} \sum_k (k-1) f_k(t) \to \frac{1}{2} \left(\frac{1}{1-t} - 1 \right)$$

Since the compensator converges to the deterministic limit, the Poisson convergence follows from arguments in Jacod and Shiryaev (1987). If we let $M_n(t)$ be the number of unicyclic components created up to time t then subtracting the duplicated edges we see its compensator converges to

$$\frac{1}{2}\left(\frac{1}{1-t} - 1 - 2t\right)$$

and the desired result follows.   ∎

Corollary 2.6.6.

$$EU(\lambda) = \begin{cases} h(\lambda) & \lambda < 1 \\ h(\lambda\rho) & \lambda > 1 \end{cases}$$

Proof. The result for $\lambda < 1$ follows from Theorem 2.6.5. Equations (2.6.8) and (2.6.4) imply that for $\lambda > 1$, $EU(\lambda) = EU(\lambda\rho)$.   ∎

Remark. Note that in contrast to duality result for tree sizes, the distribution of unicyclic cluster sizes is the same in $ER(n, \lambda\rho/n)$ and $ER(n, \lambda/n)$. The difference is due to fact that λ/n appears to the $(k-1)$st power in (2.6.1) and to the kth power in (2.6.7).

Conjecture 2.6.7. *If $\lambda > 1$ then as $n \to \infty$, $U_n(\lambda)$ converges to a Poisson distribution with mean $h(\lambda\rho)$.*

2.7 Critical Regime

In this section we will look at component sizes when $\lambda = 1 + \theta n^{-1/3}$ where $-\infty < \theta < \infty$, which corresponds to the critical regime for the random graph. We begin with a calculation that is simple and gives the right answer but is incorrect. Equation (2.6.2) and Stirling's formula tell us that the expected number of trees of size k is, for large k,

$$n\frac{k^{k-2}}{k!}\lambda^{k-1}e^{-\lambda k} \sim \frac{n}{\lambda\sqrt{2\pi}}k^{-5/2}(\lambda e^{1-\lambda})^k \qquad (2.7.1)$$

When $\lambda = 1$, $\lambda e^{1-\lambda} = 1$ so summing from $k = K$ to ∞, the expected number of tree of size $\geq K$ is

$$\sum_{k=K}^{\infty} \frac{n}{\sqrt{2\pi}}k^{-5/2} \sim \frac{2}{3\sqrt{2\pi}}nK^{-3/2}$$

This is small when $K \gg n^{2/3}$ suggesting that the largest tree components are of order $n^{2/3}$.

Having figured out what to guess, we will now go back and do the calculation carefully. For the moment λ is a general parameter value.

Lemma 2.7.1. *Let* $\alpha(\lambda) = \lambda - 1 - \log(\lambda)$. *If* $k \to \infty$ *and* $k = o(n^{3/4})$ *then the expected number of tree components of size* k

$$\gamma_{n,k}(\lambda) \equiv \binom{n}{k} k^{k-2} \left(\frac{\lambda}{n}\right)^{k-1} \left(1 - \frac{\lambda}{n}\right)^{k(n-k)+\binom{k}{2}-(k-1)} \tag{2.7.2}$$

$$\sim n \cdot \frac{k^{-5/2}}{\lambda\sqrt{2\pi}} \exp\left(-\alpha(\lambda)k + (\lambda - 1)\frac{k^2}{2n} - \frac{k^3}{6n^2}\right) \tag{2.7.3}$$

Proof. Using Stirling's formula and $k = o(n)$ to simplify the last exponent

$$\gamma_{n,k}(\lambda) \sim n \left[\prod_{j=1}^{k-1}\left(1 - \frac{j}{n}\right)\right] \cdot \frac{k^{-5/2}}{e^{-k}\sqrt{2\pi}} \cdot \lambda^{k-1} \left(1 - \frac{\lambda}{n}\right)^{kn-k^2/2}$$

Using the expansion $\log(1 - x) = -x - x^2/2 - x^3/3 - \ldots$ we see that if $k = o(n)$ then

$$\left(1 - \frac{\lambda}{n}\right)^{kn-k^2/2} \sim \exp(-\lambda k + \lambda k^2/2n)$$

while if $k = o(n^{3/4})$ we have

$$\prod_{j=1}^{k-1}\left(1 - \frac{j}{n}\right) = \exp\left(-\frac{1}{n}\sum_{j=1}^{k-1} j - \frac{1}{2n^2}\sum_{j=1}^{k-1} j^2 + O\left(\frac{k^4}{n^3}\right)\right) \sim \exp\left(-\frac{k^2}{2n} - \frac{k^3}{6n^2}\right)$$

Combining the last three calculations gives the desired formula. ∎

Taking $\lambda = 1$ in Lemma 2.7.1 we have

$$\gamma_{n,k}(\lambda) \sim \frac{nk^{-5/2}}{\sqrt{2\pi}} e^{-k^3/6n^2} \tag{2.7.4}$$

which has an extra term not found in (2.7.1), but still says that the largest tree components are of order $n^{2/3}$.

To see what happens when $\lambda = 1 + \theta n^{-1/3}$, note that

$$\alpha(\lambda) = \lambda - 1 - \log(\lambda) \quad \alpha(1) = 0$$
$$\alpha'(\lambda) = 1 - 1/\lambda \quad \alpha'(1) = 0$$
$$\alpha''(\lambda) = 1/\lambda^2 \quad \alpha''(1) = 1$$

Using Taylor series approximation, $\alpha(1 - \lambda) \approx \lambda^2/2$, so setting $1 - \lambda = -\theta/n^{1/3}$ the exponential in (2.7.3) is

$$\exp\left(-\theta^2 \frac{k}{2n^{2/3}} - \theta\frac{k^2}{2n^{4/3}} - \frac{k^3}{6n^2}\right)$$

Turning to more complex components we use the combinatorial fact, see Bollobás (2001, p. 120), that the number of graphs of complexity ℓ on k-labeled vertices, $v_{k,\ell}$ has

$$v_{k,\ell} \sim b_\ell k^{k+(3\ell-1)/2} \quad \text{as } k \to \infty$$

Recall that for trees ($\ell = -1$), $b_{-1} = 1$, while for unicyclic components, ($\ell = 0$), (2.6.6) tells us that $b_0 = (\pi/8)^{1/2}$.

Restricting our attention to $\lambda = 1$ for simplicity, the expected number of components of size k and complexity ℓ is

$$= \binom{n}{k} v_{k,\ell} \left(\frac{1}{n}\right)^{k+\ell} \left(1 - \frac{1}{n}\right)^{k(n-k)+\binom{k}{2}-(k+\ell)}$$

$$= \gamma_{n,k}(1) \cdot \frac{v_{k,\ell}}{k^{k-2}} \cdot \left(1 - \frac{1}{n}\right)^{-1-\ell}$$

$$\sim n \frac{k^{-5/2}}{\lambda\sqrt{2\pi}} e^{-(k^3/6n^2)} \cdot b_\ell k^{3/2+(3\ell)/2} \cdot n^{-1-\ell} \quad \text{as } k \to \infty \quad (2.7.5)$$

When $\ell = 0$, filling in the value of b_0 gives $(1/4)k^{-1}\exp(-k^3/6n^2)$ for large k. Again the largest unicyclic component will be of order $n^{2/3}$. If $k = n^a$ then the exponential factor tends to 1 for $a < 2/3$ and 0 for $a > 2/3$. Summing we see that

Theorem 2.7.2. *If $\lambda = 1$ the expected number of unicyclic components is*

$$\sim \frac{1}{4} \sum_{k=1}^{n^{2/3}} k^{-1} \sim \frac{1}{4} \cdot (2/3)\log n = \frac{1}{6}\log n \quad (2.7.6)$$

For $\ell \geq 1$, changing variables $k = xn^{2/3}$ (with $x > 0$ and hence k large):

$$= n \frac{x^{-5/2} n^{-5/3}}{\sqrt{2\pi}} e^{-x^3/6} \cdot b_\ell x^{3/2+(3\ell)/2} \cdot n^{1+\ell} \cdot n^{-1-\ell}$$

Collecting the powers of x and the powers of n together, and summing, the expected number of ℓ-components of size $\geq yn^{2/3}$ is

$$\frac{b_\ell}{\sqrt{2\pi}} \sum_{x \geq y, xn^{2/3} \in \mathbb{Z}} x^{-1+(3\ell/2)} e^{-x^3/6} \cdot n^{-2/3}$$

Again the largest components are $O(n^{2/3})$ but $(3\ell - 2)/2 \geq 1/2$ so setting $y = 0$ and recognizing the last expression as a Riemann sum we see that as $n \to \infty$ the expected number of ℓ-components converges to

$$\frac{b_\ell}{\sqrt{2\pi}} \int_0^\infty x^{(3\ell-2)/2} e^{-x^3/6} \, dx < \infty \quad (2.7.7)$$

The convergence of the integral tells us that most of the ℓ-components have size $\geq \epsilon n^{2/3}$.

Luczak, Pittel, and Weirman (1994) made a detailed study of Erdős–Rényi graphs in the critical regime. They proved, in particular, that the complexity of components remains bounded in probability as $n \to \infty$. To do this they had to consider components of all possible complexities. For this Bollobás's (1984) bounds were useful:

$$v_{k,\ell} \le \begin{cases} (c_1/\ell)^{1/2} k^{k+(3\ell-1)/2} & \ell \le k \\ (c_2 k)^{k+\ell} & 1 \le \ell \le \binom{k}{2} - k \end{cases}$$

This allowed them to show that the largest component of any complexity is $O(n^{2/3})$.

Even though there is a positive limiting probability for components of any complexity, they are rare. Janson (1993) showed that if one considered the family $ER(n, \lambda/n)$ with the obvious coupling then with probability ≈ 0.87 there is never more than one complex component – the giant component is highly complex. Janson, Knuth, Luczak, and Pittel (1993) later showed that this probability is exactly $5\pi/18$ and furthermore, at the critical value $\lambda = 1$ the probability of m 1-components and no more complex components is

$$\left(\frac{5}{18}\right)^m \cdot \sqrt{\frac{2}{3}} \cdot \frac{1}{(2m)!} \quad m = 0, 1, 2, \ldots$$

Their 136 page paper provides a wealth of detailed information. For more recent work on the size of the largest components at $\lambda = 1$, see Pittel (2001).

Back to the Random Walk Viewpoint. The combinatorial approach in the critical regime is made complicated by the fact that one must deal with components of arbitrarily large complexity. Thus we will return to our approach of exposing one vertex at a time. Let R_t be the removed sites at time t, A_t the active sites, and U_t the unexplored sites. In our first examination of this process we will stop at $\tau = \inf\{t : A_t = \emptyset\}$, but to make sure things last a while we will start with $|A_0|$ large. The ideas here are from Martin-Löf (1998), but we carry out the details somewhat differently.

As in our study of the giant component, we will speed up time and rescale the size of our sets to get a limit. To see what to guess, note that the combinatorial calculations suggest that the largest components are of order $n^{2/3}$. Since $R_t = t$ and R_τ is the size of the clusters containing A_0, we will scale time by $n^{2/3}$. When $\lambda = 1 + \theta n^{-1/3}$, $|A_t|$ will be almost a mean zero random walk. In this case $|A_t| - |A_0|$ will be $O(t^{1/2})$ so we will scale the number by $n^{1/3}$.

Having decided on the scaling, we compute the infinitesimal mean and variance. Letting $a_t = |A_t|$, $\Delta a_t = |A_{t+1}| - |A_t|$, and noticing that $u_t = |U_t| = n - t - a_t$, we have

$$E(\Delta a_t | \mathcal{F}_t) = -1 + (n - t - a_t)(1 + \theta n^{-1/3})/n$$

$$= -\frac{t + a_t}{n} + (\theta n^{-1/3})(1 - (t + a_t)/n)$$

$$\text{var}(\Delta a_t | \mathcal{F}_t) = (n - t - a_t)\frac{1 + \theta n^{-1/3}}{n}\left(1 - \frac{1 + \theta n^{-1/3}}{n}\right)$$

Speeding up time by $n^{2/3}$, dividing by $n^{1/3}$, and recalling $a_{\lfloor sn^{2/3} \rfloor} = O(n^{1/3})$,

$$
E\left(\frac{\Delta a_{\lfloor sn^{2/3} \rfloor}}{n^{1/3}} \,\middle|\, \mathcal{F}_{\lfloor sn^{2/3} \rfloor} \right) = \frac{-\lfloor sn^{2/3} \rfloor - a_{\lfloor sn^{2/3} \rfloor}}{n \cdot n^{1/3}} + \frac{\theta n^{-1/3} + o(n^{-1/3})}{n^{1/3}}
$$

$$
= (-s + \theta) \cdot n^{-2/3} + o(n^{-2/3})
$$

The variance is much easier

$$
\mathrm{var}\left(\frac{\Delta a_{\lfloor sn^{2/3} \rfloor}}{n^{1/3}} \,\middle|\, \mathcal{F}_{\lfloor sn^{2/3} \rfloor} \right) \sim \frac{1}{n^{2/3}}
$$

Letting $n \to \infty$ we see that $a_{\lfloor sn^{2/3} \rfloor}/n^{1/3}$ converges in distribution to the solution of

$$
da_s = (-s + \theta)\, ds + dB_s
$$

which is simply $a_s = B_s + \theta s - s^2/2$ (run until the first time it hits zero).

Aldous' Theorem. Our next goal is to describe a remarkable result of Aldous (1997) that gives the joint distribution of the sizes of the large clusters divided by $n^{2/3}$ and their complexity. The paper that proves this result is 43 pages long, so we will content ourselves to explain why this is true, making a number of Olympian leaps of faith in the process.

Consider now the version of RAU in which we choose $i_t \in U_t$ when $A_t = \emptyset$. This adds one to $|A_t|$ each time it hits zero. If we do this to a simple random walk that goes up or down by 1 with probability $1/2$ on each step, we end up with a Markov chain S_n that has $p(0, 1) = 1$ and $p(k, k-1) = p(k, k+1) = 1/2$ when $k > 0$. As $n \to \infty$, $S_n/\sqrt{n}$ converges to reflecting Brownian motion, $\hat{B}_t$, a process that is often defined as $|B_t|$, but is perhaps better seen through the eyes of Paul Lévy:

$$
\hat{B}_t = B_t - \min_{0 \le s \le t} B_s
$$

Here the second term is the number of ones we have added to keep $S_n \ge 0$. In a similar way, adding 1 each time $|A_t|$ hits zero changes the limit process to

$$
\hat{a}_t = a_t - \min_{0 \le s \le t} a_s
$$

When $|A_t| = k$, self-intersections happen at rate k/n. In our normalization $k = xn^{1/3}$ so self-intersections happen at rate $xn^{-2/3}$, or after time is sped up by a factor of $n^{2/3}$, at rate x. This motivates the introduction of a point process N_t that has points at rate $\hat{a}_t$, or more formally N_t is a counting process (i.e., nonnegative integer valued, nondecreasing, and all jumps of size one) so that

$$
N_t - \int_0^t \hat{a}_s \, ds \quad \text{is a martingale}
$$

We say that (u, v) is an excursion interval of $\hat{a}_t$ if $\hat{a}_u = \hat{a}_v = 0$ but $\hat{a}_t \ne 0$ for $t \in (u, v)$. During excursions $|A_t|$ does not hit 0. Since we expose one vertex at a time, the lengths of the excursion intervals represent cluster sizes in the random graph. $N_v - N_u$ gives the number of self-intersections that happened during the interval. Following Aldous' terminology we will call this the surplus, that is, the number of extra edges compared to a tree of the same size (= the complexity + 1).

Theorem 2.7.3. *Let* $K_1^n \geq K_2^n \geq \ldots$ *be the ordered component sizes of* $ER(n, (1 + \theta n^{-1/3})/n)$ *and let* $o_1^n, o_2^n, o_3^n \ldots$ *be the corresponding surpluses. Then as* $n \to \infty$, $\{(n^{-2/3}K_j^n, \sigma_j^n) : j \geq 1\}$ *converges in distribution to* $\{(L_j, \sigma_j) : j \geq 1\}$ *where* $L_1 > L_2 > L_3 > \ldots$ *are the ordered lengths of excursion intervals in* $\{\hat{a}_s : s \geq 0\}$ *and* $\sigma_1, \sigma_2, \sigma_3, \ldots$ *are the number of N-arrivals in these intervals.*

The limit distribution is not very explicit, but it does tell us that the largest cluster is $O(n^{2/3})$ and that the number of ℓ-components converges in distribution.

Multiplicative Coalescent. While Theorem 2.7.3 is nice, the truly remarkable part of Aldous' contribution is to view the large clusters in $ER(n, (1 + tn^{-1/3})/n)$ as a process indexed by $-\infty < t < \infty$. The first detail is to construct all of the random graphs on the same space but this is easy: we assign independent random variables ζ_e uniformly distributed on $(0, 1)$ to each edge and declare that the edge is present if $\zeta_e < (1 + tn^{-1/3})/n$. The state at time t will be the ordered component sizes $\{K_j^n(t)/n^{2/3}\}$. Consider two clusters of sizes $xn^{2/3}$ and $yn^{2/3}$. In a short interval of time $(t, t + h)$, an individual edge is added with probability $hn^{-4/3}$, so the probability of making a connection between the two clusters is $\approx xyh$.

Thus in the limit as $n \to \infty$ we expect to get a limit process that follows the simple rule: each pair of clusters of sizes (x, y) merges at rate xy to a cluster of size $x + y$. To have an honest Markov process, we need a state space. Aldous chose $\ell_{\searrow}^2$ the collection of decreasing sequences $x_1 \geq x_2 \geq x_3 \geq \ldots$ with $\sum_k x_k^2 < \infty$. The other thing that should be noticed is that the time interval is $-\infty < t < \infty$ so there is no initial distribution. In Aldous' original paper he solved this problem by showing that there was only one "standard multiplicative coalescent" that had the one-dimensional distributions consistent with Theorem 2.7.3. Aldous and Limic (1998) later characterized all processes on $-\infty < t < \infty$ in which each pair of clusters of sizes (x, y) merges at rate xy to a cluster of size $x + y$. (In terms of the theory of Markov processes one is finding all of the entrance laws.) In addition to the constant process $(x_1 = v > 0, x_i = 0, i \geq 2)$ there are some nonstandard ones, which are irrelevant for the following application.

Theorem 2.7.4. *Let* $K_1^n(t) \geq K_2^n(t) \geq \ldots$ *be the ordered component sizes of* $ER(n, (1 + tn^{-1/3})/n)$. *As* $n \to \infty$, $\{K_j^n(t)/n^{2/3} : j \geq 1\}$, $-\infty < t < \infty$ *converges in distribution to the standard multiplicative coalescent.*

The convergence of rescaled large components to the multiplicative coalescent, provides a nice intuitive process of the growth of clusters in the critical regime. Alon and Spencer (2002) describe the evolution as follows: "With $t = -10^6$, say we have feudalism. Many components (castles) are each vying to be the largest. As t increases the components increase in size and a few large components (nations) emerge. An already large France has much better chances of becoming larger than a smaller Andorra. The largest components tend to merge and by $t = 10^6$ it is very likely that a giant component, the Roman Empire, has emerged. With high probability this component is nevermore challenged for supremacy but continues absorbing smaller components until full connectivity – One World – is achieved."

Remark. Yuval Peres has used ideas from Martin-Löf (1998) to show that if $n > 7$ then the largest component $\mathcal{C}^{(1)}$ has

$$P(|\mathcal{C}^{(1)}| > An^{2/3}) \leq \frac{3 + 6n^{-1/3}}{A^2}$$

The key idea is to upper bound Δa_t by $-1 + \text{Binomial}(n, 1/n)$ for $t < n^{2/3}$ and by $-1 + \text{Binomial}(n - n^{2/3}, 1/n)$ for $t \geq n^{2/3}$. Nachmias and Peres (2005) prove in addition an upper bound on $P(|\mathcal{C}^{(1)}| < \delta n^{2/3})$ for δ small.

2.8 Threshold for Connectivity

In this section we will investigate the question: How large does λ have to be so that the probability $ER(n, \lambda/n)$ is connected (i.e., ALL vertices in ONE component) tends to 1. Half of the answer is easy. Let d_x be the degree of x.

$$P(d_x = 0) = \left(1 - \frac{\lambda}{n}\right)^{n-1}$$

Using the series expansion $\log(1 - x) = -x - x^2/2 - x^3/3 - \ldots$ it is easy to see that if $\lambda = o(n^{1/2})$ then

$$n \log\left(1 - \frac{\lambda}{n}\right) = -\lambda - \lambda^2/2n - \lambda^3/3n^2 - \ldots$$

and hence

$$\left(1 - \frac{\lambda}{n}\right)^n e^\lambda \to 1 \tag{2.8.1}$$

Thus when $\lambda = a \log n$ we have $P(d_x = 0) \sim n^{-a}$, and if $a < 1$ the number of isolated vertices $I_n = |\{x : d_x = 0\}|$ has

$$EI_n = nP(d_x = 0) \sim n^{1-a} \to \infty$$

To show that the actual value of I_n is close to the mean we note that if $x \neq y$

$$P(d_x = 0, d_y = 0) = \left(1 - \frac{\lambda}{n}\right)^{2n-3} = \left(1 - \frac{\lambda}{n}\right)^{-1} P(d_x = 0)P(d_y = 0)$$

so we have

$$\text{var}(I_n) = nP(d_1 = 0)(1 - P(d_1 = 0))$$
$$+ n(n-1)\left(\left(1 - \frac{\lambda}{n}\right)^{-1} - 1\right) P(d_1 = 0)P(d_2 = 0)$$

When $\lambda = a \log n$

$$\text{var}(I_n) \sim n^{1-a} + n^2 \left(\frac{\lambda}{n}\right) n^{-2a} \sim EI_n$$

Using Chebyshev's inequality it follows that if $a < 1$

$$P\left(|I_n - EI_n| > (\log n)(EI_n)^{1/2}\right) \leq \frac{1}{\log^2 n} \qquad (2.8.2)$$

The last result shows that if $\lambda = a \log n$ with $a < 1$ then with high probability there are about n^{1-a} isolated vertices, and hence the graph is not connected. Showing that the graph is connected is more complicated because we have to consider all possible ways in which the graph can fail to be connected. Equation (2.6.2) tells us that the expected number of trees of size k is

$$\sim n \frac{k^{k-2}}{k!} \lambda^{k-1} e^{-\lambda k}$$

When $k = 2$ and $\lambda = a \log n$ this is

$$\frac{n}{2}(a \log n)n^{-2a}$$

Thus if $1/2 < a < 1$ there are isolated vertices, but no components of size 2. It is easy to generalize the last argument to conclude that when $\lambda = a \log n$ and $1/(k+1) < a < 1/k$ there are trees of size k but not of size $k + 1$. Bollobás (2001), see Section 7.1, uses this observation with the fact we know that the largest component is $O(\log n)$ to sum the expected values and prove:

Theorem 2.8.1. *Consider $G = ER(n, \lambda/n)$ with $\lambda = a \log n$. The probability G is connected tends to 0 if $a < 1$ and to 1 if $a > 1$.*

Proof. We will use the approach of Section 2.3 to show that the probability a vertex fails to connect to the giant component is $o(1/n)$. Since we have a large λ we can use the lower bound process with $\delta = 1/2$. The constant $\theta_{1/2}$ that appears in (2.3.7) is defined by

$$\theta + (\lambda/2)(e^{-\theta} - 1) = 0$$

This is hard to compute for fixed λ, so instead we decide we want $\theta = 1$ and see that this means $\lambda = 2e/(e-1)$. Using monotonicity we see that if $\lambda \geq 2e/(e-1)$, (2.3.8) implies that for our comparison random walk

$$P_{2\log n}(T_0 < \infty) \leq n^{-2}$$

To reach size $2 \log n$ we use the large deviations bound in Lemma 2.3.3, with $\delta = 0$ and $x = 1/2$ to conclude that if S_K is a sum of K independent Binomial$(n, \lambda/n)$ random variables then $\mu = K\lambda$ and

$$P(S_K \leq \mu/2) \leq \exp(-\gamma(1/2)\mu) \tag{2.8.3}$$

where $\gamma(1/2) = (1/2)\log(1/2) - 1/2 + 1 = 1/2(1 - \log(2)) \geq 0.15$. If $\lambda = (1 + \epsilon) \log n$ with $\epsilon \geq 0$ and $K = 14$ then

$$P(S_{14} \leq 7 \log n) \leq n^{-2.1}$$

The last calculation shows that if x has at least 14 neighbors, then with high probability at distance 2 there are at least $7 \log n$ vertices. The next step is to bound

$$\begin{aligned}
P(d_x \leq 13) &= \sum_{k=0}^{13} \binom{n}{k} \left(\frac{\lambda}{n}\right)^k \left(1 - \frac{\lambda}{n}\right)^{n-k} \\
&\leq \sum_{k=0}^{13} \frac{\lambda^k}{k!} e^{-\lambda(n-k)/n} \leq 14(a \log n)^{13} n^{-a} e^{(13a \log n)/n}
\end{aligned}$$

which is $o(n^{-1})$ if $a > 1$.

To finish up now (and to prepare for the next proof), we apply the large deviations result Lemma 2.3.3 to lower bounding random walk W_t twice to conclude that if $-1 + (a \log n) \geq (a/2) \log n$ then there are positive constants η_i so that

$$P(W(n^{1/2}) - W(0) \leq (a/2)n^{1/2} \log n) \leq \exp(-\eta_1 n^{1/2} \log n)$$
$$P(W(n^{1/2}) + n^{1/2} - W(0) \geq 2an^{1/2} \log n) \leq \exp(-\eta_2 n^{1/2} \log n)$$

Combining our results we see that with probability $1 - o(n^{-1})$ the RAU process will not expose $n/2$ vertices and have at least $0.1n^{1/2} \log n$ active vertices at time $n^{1/2}$. When this occurs for x and for y the probability that their clusters fail to intersect is at most

$$\left(1 - \frac{\log n}{n}\right)^{0.01n(\log n)^2} \leq e^{-(\log n)^3/100}$$

and the proof is complete. ∎

The next result is an easy extension of the previous argument and will allow us to get a sharper result about the transition to connectivity.

Theorem 2.8.2. *Consider $G = ER(n, \lambda/n)$ with $\lambda = a \log n$. If $a > 1/2$ then with probability tending to 1, G consists only of a giant component and isolated vertices.*

Proof. It follows from the previous argument that if n is large $P_{2 \log n}(T_0 < \infty) \leq n^{-2}$. Using (2.8.3) it is easy to see that if S_{28} is a sum of 28 independent Binomial$(n, \lambda/n)$ and $\lambda \geq (1/2) \log n$ then

$$P(S_{28} \leq 7 \log n) \leq n^{-2.1}$$

Consider now a branching process Z_k^x with offspring distribution Binomial $(n, \lambda/n)$. If $Z_1^x = 0$ the cluster containing x is a singleton. We might be unlucky and have $Z_1^x = 1$ but in this case

$$P(Z_1^x = 1, Z_2^x \leq 27) = n \frac{\lambda}{n} \left(1 - \frac{\lambda}{n}\right)^{n-1} \sum_{k=0}^{27} \binom{n}{k} \left(\frac{\lambda}{n}\right)^k \left(1 - \frac{\lambda}{n}\right)^{n-k}$$

$$\leq \sum_{k=0}^{27} \frac{\lambda^{k+1}}{k!} e^{-\lambda(2n-k-1)/n} \leq 28(a \log n)^{28} n^{-2a} e^{(28a \log n)/n}$$

which is $o(n^{-1})$ if $a > 1/2$. If we are lucky enough to find two neighbors on the first try then

$$P(Z_2^x \leq 28 | Z_1^x \geq 2) \leq \sum_{k=0}^{28} \binom{2n}{k} \left(\frac{\lambda}{n}\right)^k \left(1 - \frac{\lambda}{n}\right)^{2n-k}$$

$$\leq \sum_{k=0}^{28} \frac{(2\lambda)^k}{k!} e^{-2\lambda(n-k)/n} \leq 29(a \log n)^{28} n^{-2a} e^{(56a \log n)/n}$$

Here we have replaced 27 by 28 to take care of collisions. One collision has probability c/n, but by a now familiar estimate, the probability of two collisions in one step of the branching process is $O(1/n^2)$.

In the previous proof we only used $a > 1/2$ in the W-estimates and the final estimate so the proof is complete. ∎

We are now ready to more precisely locate the connectivity threshold.

Theorem 2.8.3. *Consider $G = ER(n, \lambda/n)$ with $\lambda = \log n + b + o(1)$. The number of isolated vertices I_n converges to a Poisson distribution with mean e^{-b} and hence the probability G is connected tends to $\exp(-e^{-b})$.*

Proof. By the previous result G will be connected if and only if there are no isolated vertices. Using (2.8.1), the probability x is isolated is

$$\left(1 - \frac{\lambda}{n}\right)^n \sim \exp(-\log n - b) \sim e^{-b}/n$$

so $E I_n$, the expected number of ordered k-tuples of isolated vertices, is

$$(n \cdot (n-1) \cdots (n-k+1)) \left(1 - \frac{\lambda}{n}\right)^{n+(n-1)+\cdots+(n-k+1)} \to e^{-bk}$$

so the Poisson convergence follows from the method of moments. $\blacksquare$

Having shown that the probability $E R(n, c(\log n)/n)$ is connected tends to 1 for $c > 1$, our next question is to determine its diameter. The heuristic is the same as in Theorem 2.4.1 for $E R(n, c/n)$. We set $(np)^m = n$ and solve to conclude that the average distance between two points is $\log n/(\log np)$. For $E R(n, c(\log n)/n)$ this is also the diameter. We begin with a large deviations result for the Binomial distribution.

Lemma 2.8.4. *Let* $X = Binomial(n, p)$ *and let* $H(a) = a \log(a/p) + (1-a) \log((1-a)/(1-p))$. *Then for* $b < p < c$

$$P(X \leq nb) \leq \exp(-H(b)n) \qquad P(X \geq nc) \leq \exp(-H(c)n)$$

Remark. To see why this is the answer, note that from the definition of the Binomial distribution and Stirling's formula, $n! \sim n^n e^{-n}\sqrt{2\pi n}$,

$$P(X = na) = \binom{n}{na} p^{na}(1-p)^{n(1-a)} \approx \frac{n^n}{(na)^{na}(n(1-a))^{n(1-a)}} p^{na}(1-p)^{n(1-a)}$$

Now cancel the n^n's in the fraction and take $(1/n)\log$ of both sides.

Proof. Suppose $a > p$. Then for any $\theta > 0$

$$e^{na\theta} P(X \geq na) \leq ((1-p) + pe^{\theta})^n$$

which we can rewrite as

$$P(X \geq na) \leq \left[\exp\{-\theta a + \log((1-p) + pe^{\theta})\}\right]^n$$

To optimize we differentiate the term in set braces with respect to θ and set

$$0 = -a + \frac{pe^{\theta}}{(1-p) + pe^{\theta}}$$

Solving we have $e^\theta = (a(1-p))/(p(1-a))$. Using $((1-p)+pe^\theta) = pe^\theta/a = (1-p)/(1-a)$ and plugging in we have

$$= \exp\left(-a\log\left(\frac{a(1-p)}{p(1-a)}\right) + \log((1-p)/(1-a))\right) = \exp(-H(a))$$

For $a < p$ the first inequality is valid for $\theta < 0$ but the remaining calculations are the same. ∎

In many cases the following simplification is useful.

Lemma 2.8.5. *If $X = Binomial(n, p)$ then*

$$P(X \le n(p-z)) \le \exp(-nz^2/2p) \qquad P(X \ge n(p+z)) \le \exp(-nz^2/2(p+z))$$

Taking $z = py$ this becomes

$$P(X \le np(1-y)) \le \exp(-npy^2/2) \qquad P(X \ge np(1+y)) \le \exp(-npy^2/2(1+y))$$

Remark. If X/n were Gaussian with variance $p(1-p)/n$ these probabilities would be $\le \exp(-nz^2/2p(1-p))$.

Proof. We begin with the second inequality. The function defined in Lemma 2.8.4 has $H(p) = 0$ and

$$H'(a) = \log(a/p) - \log((1-a)/(1-p))$$

so $H'(p) = 0$. If $p < a \le 1$, Taylor's theorem implies that there is a $y \in [p, a]$ so that $H(a) = H''(y)(a-p)^2/2$. Differentiating again $H''(a) = 1/a(1-a)$ which is minimized at $1/2$. Therefore if $a \le 1/2$, $H(a) \ge (a-p)^2/2a$, while if $a \ge 1/2$, $H(a) \ge 2(a-p)^2 \ge (a-p)^2/2a$. Substituting $a = p+z$ gives the second result. The argument for the first result is almost identical, but when we substitute $a = p-z$ we can let $a = p$ in the denominator. ∎

Theorem 2.8.6. *If $\liminf np/(\log n) > c > 1$ and $(\log np)/(\log n) \to 0$ then the diameter of $ER(n, p)$, $D(n, p) \sim (\log n)/(\log np)$.*

Remark. The first condition guarantees that the probability that the graph is connected tends to 1 as $n \to \infty$. To explain the second suppose $p = n^{15/17}$. In this case $np = n^{2/17}$ and it is not hard to show that the diameter will be 9 with probability approaching 1, but the formula gives $17/2$. A more delicate situation occurs when

$$p = n^{(1/d)-1}(\log(n^2/c))^{1/d}$$

In this case in the limit the diameter is d with probability $e^{-c/2}$ and $d + 1$ with probability $1 - e^{-c/2}$. See Theorem 10.10 in Bollobás (2001). Chung and Lu (2001) have shown, see Theorem 5 on page 272, that the conclusion of Theorem 2.8.6 holds for the giant component if $\log n > np \to \infty$.

Proof. Fix x and let $B_k(x) = \{y : d(x, y) = k\}$. $|B_k|$ is dominated by Z_k a branching process with a Binomial(n, p) offspring distribution. Using Lemma 2.8.5 with $y = 3$ we have

$$P(Z_1 \geq 4np) \leq \exp(-9np/8) \leq n^{-9c/8}$$

for large n. Using Lemma 2.8.5 again with n replaced by $(k + 2)(np)^{k-1}n$ and $y = 1/(k + 2)$ we have

$$P(Z_k > (k + 3)(np)^k | Z_{k-1} \leq (k + 2)(np)^{k-1}) \leq \exp(-(np)^k/(k + 3))$$

Since $np \geq c \log n$ for large n the right-hand side converges to 0 very rapidly as $n \to \infty$ and we have

$$P(Z_k \leq (k + 3)(np)^k \quad \text{for} \quad 1 \leq k \leq (\log n)/(\log np)) = 1 - O(n^{-9c/8})$$
$$(2.8.4)$$

To get a lower bound we note that by the proof of Theorem 2.8.1, $P(|B_1| \leq 13) = 1 - o(n^{-1})$ and

$$P(|B_2| \leq 7np \,|\, |B_1| \geq 14) \leq n^{-2.1}$$

To control the growth of $|B_k|$ for $k \geq 3$ we note that as long as $\sum_{j=0}^k |B_j| \leq n^{2/3}$, $|B_k|$ dominates a branching process with a Binomial (n', p) offspring distribution where $n' = n - n^{2/3}$ and $n'p \geq c \log n$ for large n. Define a sequence of constants by $a_2 = 7$ and for $k \geq 3$, $a_k = a_{k-1}(1 - 1/k^2)$. Since a_k is decreasing, at each iteration we subtract less than $7/k^2$. $\sum_{k=3}^{\infty} 1/k^2 = \pi^2/6 - 1 - 1/4 \leq 0.4$ so $a_k \geq 4$ for all k. Using Lemma 2.8.5 again with n replaced by $a_{k-1}(n'p)^{k-2}n'$ and $y = 1/k^2$ we have for $k \geq 3$

$$P(|B_k| < a_k(n'p)^{k-1} | Z_{k-1} \geq a_{k-1}(n'p)^{k-2}) \leq \exp(-(n'p)^{k-1}/2k^4)$$

Since $n'p \geq c \log n$ for large n the right-hand side converges to 0 very rapidly as $n \to \infty$ and we have

$$P(|B_k| \geq 4(n'p)^{k-1} \quad \text{for} \quad 2 \leq k \leq (0.6)(\log n)/(\log np)) = 1 - o(n^{-1})$$
$$(2.8.5)$$

Here 0.6 is chosen so that (2.8.4) implies that for large n we have $\sum_{j=0}^k |B_j| \leq n^{2/3}$ the indicated range of k's.

The bound in (2.8.4) implies that if $\epsilon > 0$ then with probability $1 - O(n^{-9c/8})$

$$\sum_{k=0}^{(1-\epsilon)\log n/(\log np)} Z_k < n$$

for large n so $\liminf D(n, p)/(\log n/(\log np)) \geq 1 - \epsilon$. Let k be the first integer larger than $1 + ((1 + \epsilon)(\log n))/(2(\log np))$. Using (2.8.5) we see that if $\epsilon > 0$ then

$$P(|B_k| \geq 4n^{(1+\epsilon)/2}) = 1 - o(n^{-1})$$

By a now familiar estimate, when this occurs for two different starting points then either the two clusters have already intersected or with probability $\geq \exp(-16n^{(1+\epsilon)})$ they will do so on the next step. This shows that $\limsup D(n, p)/(\log n/(\log np)) \leq 1 + \epsilon$ and completes the proof. ∎

3

Fixed Degree Distributions

3.1 Definitions and Heuristics

In an Erdös–Rényi random graph, vertices have degrees that have asymptotically a Poisson distribution. However, as discussed in Section 1.4, in social and communication networks, the distribution of degrees is much different from the Poisson and in many cases has a power law form, that is, the fraction of vertices of degree k, $p_k \sim Ck^{-\beta}$ as $k \to \infty$. Molloy and Reed (1995) were the first to construct graphs with specified degree distributions. We will use the approach of Newman, Strogatz, and Watts (2001, 2002) to define the model.

Let $d_1, \ldots d_n$ be independent and have $P(d_i = k) = p_k$. Since we want d_i to be the degree of vertex i, we condition on $E_n = \{d_1 + \cdots + d_n \text{ is even}\}$. If the probability $P(E_1) \in (0, 1)$ then $P(E_n) \to 1/2$ as $n \to \infty$ so the conditioning will have little effect on the finite-dimensional distributions. If d_1 is always even then $P(E_n) = 1$ for all n, while if d_1 is always odd, $P(E_{2n}) = 1$ and $P(E_{2n+1}) = 0$ for all n.

To build the graph we think of d_i half-edges attached to i and then pair the half-edges at random. The picture gives an example with eight vertices.

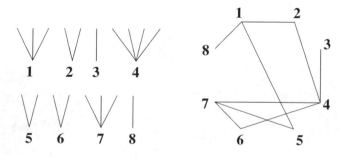

This construction can produce multiple edges and even self-loops but there are not very many, if we suppose, as we will in the first four sections of this chapter, that

(A) p_k has a finite second moment.

Thus the probability that the graph is simple has a positive limit. We begin with a useful result.

Lemma 3.1.1. *Let* $x^{(k)} = x(x-1)\cdots(x-k+1)$ *and let* Z_n *be a sequence of nonnegative integer-valued random variables with* $EZ_n^{(k)} \to \lambda^k$ *for all positive integers k. Then* Z_n *converges in distribution to* $Z_\infty = Poisson(\lambda)$.

Proof. The factorial moments of the Poisson $EZ_\infty^{(k)} = \lambda^k$. Since $x^k = \sum_{j=1}^k c_{k,j} x^{(j)}$ we have $EZ_n^k \to EZ_\infty^k$. The Poisson distribution is determined by its moments so the conclusion follows, see for example, Section 2.3.e in Durrett (2004). ∎

Theorem 3.1.2. *Let* $\mu = \sum_k k p_k$ *and* $\mu_2 = \sum_k k(k-1)p_k$. *As* $n \to \infty$, *the number of self-loops* χ_0 *and the number of parallel edges* χ_1 *are asymptotically independent* $Poisson(\mu_2/2\mu)$ *and* $Poisson((\mu_2/2\mu)^2)$.

Proof. If $D = d_1 + \cdots + d_n$ then the expected number of self-loops is $\sim \sum_i d_i(d_i - 1)/2D \to \mu_2/2\mu$ and the expected number of pairs of parallel edges is

$$\sim \frac{1}{2} \sum_i \frac{d_i d_{i-1}}{2} \sum_{j \neq i} \frac{d_j}{D} \frac{d_j - 1}{D} \to (\mu_2/2\mu)^2 0$$

To check the counting in the second formula, suppose that we will connect half-edges 3 and 7 from i to half-edges 5 and 2 from j. When we pick the two edges for i, order is not important, but when it comes to j it is. The 1/2 in front comes from the fact that (i, j) and (j, i) are both in the double sum.

This gives the asymptotics for the mean, but in almost the same way we can compute the expected number of ordered k-tuples (for more details in a similar situation see the proof of Theorem 2.4.2) to conclude that the self-loops and parallel edges have Poisson limits. After this, we can consider product moments to get convergence of the joint distribution to independent Poissons. ∎

As we have seen in our analysis of Erdös–Rényi graphs, the growth of clusters can be approximated in the early stages by a branching process. In this section we will follow Newman, Strogatz, and Watts (2001) who were playing by the rules of *Physical Review E*. We will assume that the growth of the cluster is a branching process, ignore the technicalities of estimating the difference, and compute the answer. If we start with a given vertex x then the number of neighbors (the first generation in the branching process) has distribution p_j. This is not true for the second generation. A first generation vertex with degree k is k times as likely to be chosen as one with degree 1, so the distribution of the number of children of a first generation vertex is for $k \geq 1$

$$q_{k-1} = \frac{kp_k}{\mu} \quad \text{where} \quad \mu = \sum_k kp_k$$

The $k - 1$ on the left-hand side comes from the fact that we used up one edge connecting to the vertex. Note that since we have assumed p has finite second moment, q has finite mean $\nu = \sum_k k(k-1)p_k/\mu$.

For a concrete example, consider the Poisson distribution $p_k = e^{-\lambda}\lambda^k/k!$, which is the asymptotic degree distribution for $ER(n, \lambda/n)$. In this case $\mu = \lambda$, so

$$q_{k-1} = e^{-\lambda}\frac{k\lambda^k}{k!\lambda} = e^{-\lambda}\frac{\lambda^{k-1}}{(k-1)!}$$

and q is again Poisson. Conversely, if $p = q$ we have $p_k = p_{k-1}\mu/k$. Iterating gives $p_k = p_0\mu^k/k!$, so p_k is Poisson with mean μ.

Having discovered that the growth of the cluster is a "two-phase" branching process in which the first generation has distribution p, and the second and subsequent generations have distribution q, it is now routine to compute the threshold for the existence of large components and the size of such components. Let $G_0(z) = \sum_k p_k z^k$ and $G_1(z) = \sum_k q_k z^k$ be the two generating functions. These are related:

$$\frac{G_0'(z)}{\mu} = \sum_{k=1}^{\infty} \frac{kp_k}{\mu}z^{k-1} = \sum_{k=1}^{\infty} q_{k-1}z^{k-1} = G_1(z)$$

Let Z_n be the number of vertices in the nth generation. For $n \geq 1$, $EZ_n = \mu\nu^{n-1}$ so when $\nu < 1$

$$E\left(\sum_{n=0}^{\infty} Z_n\right) = 1 + \sum_{n=1}^{\infty} \mu\nu^{n-1} = 1 + \frac{\mu}{1-\nu}$$

and we have the first part of

Theorem 3.1.3. *The condition for the existence of a giant component is $\nu > 1$. In this case the fraction of vertices in the giant component is asymptotically $1 - G_0(\rho_1)$ where ρ_1 is the smallest fixed point of G_1 in $[0, 1]$.*

We will prove this result in the next section. To explain the second claim: We know that ρ_1 is the extinction probability of the homogeneous branching process with offspring distribution q. Breaking things down according to the number of offspring in the first generation, the probability that the two phase branching process dies out is $\sum_{k=0}^{\infty} p_k\rho_1^k = G_0(\rho_1)$, since each of the k first generation families must die out and they are independent.

In the remainder of this section we will analyze the model as Newman, Strogatz, and Watts did, which means that we will rely on generating functions. To compute the distribution of the size of the nongiant components, we begin by considering the homogeneous branching process with offspring distribution q. Let $H_1(z)$ be the generating function of the total progeny in the homogeneous branching process. If there are k offspring in the first generation then the total progeny $= 1 + Y_1 + \cdots + Y_k$ where the Y_i are independent and have the same distribution as the total progeny. From this it follows that for $z < 1$

$$H_1(z) = z \cdot \sum_{k=0}^{\infty} q_k(H_1(z))^k = zG_1(H_1(z)) \qquad (3.1.1)$$

The condition $z < 1$ guarantees $z^{\infty} = 0$. $H_1(0) = 0$ reflecting the fact that the total progeny is always at least one. We claim that $H_1(1) = \rho_1$. To check this, note that when $z < 1$, $G_1(H_1(z))/H_1(z) = 1/z > 1$ so $H_1(z) < \rho_1$, but $\lim_{z \to 1} H_1(z)$ is a fixed point of G_1 so the limit must be ρ_1. Similar reasoning shows that if $H_0(z)$ be the generating function of the total progeny in the branching process then

$$H_0(z) = z \cdot \sum_{k=0}^{\infty} p_k(H_1(z))^k = zG_0(H_1(z)). \qquad (3.1.2)$$

Mean Size of Finite Clusters. We begin with the homogeneous branching process. Differentiating (3.1.1), setting $z = 1$ and recalling $H_1(1) = \rho_1$, $G_1(\rho_1) = \rho_1$,

$$H_1'(z) = G_1(H_1(z)) + zG_1'(H_1(z))H_1'(z)$$
$$H_1'(1) = \rho_1 + G_1'(\rho_1)H_1'(1)$$

Rearranging we have

$$\frac{H_1'(1)}{H_1(1)} = \frac{1}{1 - G_1'(\rho_1)} \qquad (3.1.3)$$

When the branching process dies out, $\rho_1 = 1$, $G_1'(1) = \nu$ and $H_1(1) = 1$, and so this says that the mean cluster size is $1/(1 - \nu)$. To check the result for $\nu > 1$, note that the proof of Theorem 2.1.8 implies that if the homogeneous branching process is supercritical and we condition it to die out then the result is a branching process with offspring distribution that has generating function $G_1(\rho_1 z)/\rho_1$. Differentiating and setting $z = 1$, we see that the mean of the offspring distribution is $G_1'(\rho_1)$, so the expected total progeny will be $\sum_{n=0}^{\infty} G_1'(\rho_1)^n = 1/(1 - G_1'(\rho_1))$.

Turning to the two-phase branching process. Differentiating (3.1.2) and setting $z = 1$

$$H_0'(z) = G_0(H_1(z)) + zG_0'(H_1(z))H_1'(z)$$
$$H_0'(1) = H_0(1) + G_0'(\rho_1)H_1'(1)$$

since $H_1(1) = \rho_1$ and $G_0(\rho_1) = H_0(1)$ is the extinction probability for the two phase process. When the branching process dies out, $H_0(1) = 1$, $\rho_1 = 1$, $G_0'(1) = \mu$,

$H_1(1) = 1$, and $H_1'(1) = 1/(1 - \nu)$ so we have

$$H_0'(1) = 1 + \frac{\mu}{1 - \nu}$$

as we have computed previously. To get a result for $\nu > 1$, we divide each side by $H_0(1)$ and use (3.1.3) to get

$$\frac{H_0'(1)}{H_0(1)} = 1 + \frac{G_0'(\rho_1)H_1'(1)}{H_0(1)} = 1 + \frac{G_0'(\rho_1)H_1(1)}{H_0(1)(1 - G_1'(\rho_1))} \qquad (3.1.4)$$

To check this, note that in the two-phase branching process the probability $Z_1 = k$ given that the process dies out is $p_k \rho_1^k / G_0(\rho_1)$. Equation (3.1.3) implies that the mean number of descendants of a first generation individual given the extinction of its family is $1/(1 - G_1'(\rho_1))$. Thus the expected total progeny is

$$1 + \sum_{k=1}^{\infty} \frac{p_k \rho_1^k}{G_0(\rho_1)} \frac{k}{1 - G_1'(\rho_1)} = 1 + \frac{\rho_1 G_0'(\rho_1)}{G_0(\rho_1)} \cdot \frac{1}{1 - G_1'(\rho_1)}$$

Since $H_1(1) = \rho_1$ and $G_0(\rho_1) = H_0(1)$ this agrees with the answer in (3.1.4).

Cluster Size at the Critical Value. Taking $w = H_1(z)$ in (3.1.1) gives

$$z = H_1^{-1}(w) = w/G_1(w)$$

In the critical case $G_1(1) = 1$ and $G_1'(1) = 1$. Taking the derivative of the right-hand side and setting $w = 1$ is

$$\frac{G_1(w) - G_1'(w)w}{G_1(w)^2} = 0$$

Differentiating again and setting $w = 1$ gives

$$\frac{[G_1'(w) - G_1''(w)w - G_1'(w)]G_1(w)^2 - 2G_1(w)G_1'(w)[G_1(w) - G_1'(w)w]}{G_1(w)^4} = -G_1''(1)$$

If q has a finite second moment then for some $x_\epsilon \in [1 - \epsilon, 1]$

$$H_1^{-1}(1 - \epsilon) = 1 - \frac{1}{2}G_1''(x_\epsilon)\epsilon^2$$

$G_1''(x_\epsilon) \to G_1''(1)$ as $\epsilon \to 0$, so as $\delta \to 0$

$$1 - H_1(1 - \delta) \sim \sqrt{2\delta/G_1''(1)}$$

Combining this with (3.1.2) we have

$$1 - H_0(1 - \delta) = 1 - (1 - \delta) + (1 - \delta)[G_0(1) - G_0(H_1(1 - \delta))]$$
$$\sim G_0'(1)\sqrt{2\delta/G_1''(1)} \qquad (3.1.5)$$

Now the rate of convergence of $H_0(1 - \delta)$ to 1 tells us about the rate of convergence of $\sum_{k=K}^{\infty} h_k$ to 0 where h_k is the probability that the total progeny is k. Tauberian theorems (see e.g., Feller (1971)) are the machinery for doing these results rigorously. Here we will be content to argue informally. Suppose that $h_k \sim Ck^{-\alpha}$.

$$1 - H_0(1 - \delta) \sim \sum_k Ck^{-\alpha}\{1 - (1 - \delta)^k\}$$

Letting $x = k\delta$ the above is

$$\sum_{x:x/\delta \in \mathbb{Z}^+} C(x/\delta)^{-\alpha}\{1 - (1 - \delta)^{x/\delta}\} \sim \delta^{\alpha-1} \cdot \delta \sum_{x:x/\delta \in \mathbb{Z}^+} Cx^{-\alpha}(1 - e^{-x})$$
$$\sim C\delta^{\alpha-1} \int_0^{\infty} x^{-\alpha}(1 - e^{-x})\,dx$$

Comparing with (3.1.5) suggests that $\alpha = 3/2$.

While the calculations above have been for the branching process, they work remarkably well for the random graph. In Section 2.6, we showed that in $ER(n, 1/n)$ the expected number of trees of size k is

$$\sim \frac{nk^{-5/2}}{\sqrt{2\pi}} e^{-k^3/3n^2}$$

Since a tree of size k contains k sites, this shows that the probability of belonging to a component of size k is asymptotically $Ck^{-3/2}$ while $k \ll n^{2/3}$. When $k = n^{2/3}$ this probability is $O(1/n)$ and there is an exponential cutoff.

3.2 Proof of Phase Transition

Again, while the branching process picture is intuitive, in order to analyze the growth of a cluster, it is more convenient to expose one vertex at a time. In the case of Erdős–Rényi random graphs, the main difference between cluster growth and a random walk is that the size of the unexplored set decreases over time. In our new setting there are two additional differences: (i) For fixed n, the empirical sequence of degrees $d_1, \ldots, d_n$ is not the same as the degree distribution; (ii) The set of available degrees changes as choices are made.

The first issue is not a problem. As $n \to \infty$ the empirical distribution of d_m, $1 \le m \le n$ converges to p_k in distribution and in total variation norm. The strong law of large numbers implies that the empirical mean of the degree distribution

$$\bar{\mu}_n = \frac{1}{n}\sum_{m=1}^{n} d_m \to \mu = \sum_k kp_k$$

The size-biased empirical distribution has

$$\bar{q}^n_{k-1} = k|\{1 \le x \le n : d_x = k\}|/(n\bar{\mu}_n)$$

and also converges to q_{k-1} in distribution and in total variation norm. Since we have supposed p_k has finite second moments, the empirical mean

$$\bar{\nu}_n = \sum_k k\bar{q}^n_k \to \nu$$

To deal with (ii), we will only look at cluster growth until a small fraction of the vertices have been exposed, and argue that the distribution of choices has not been changed much.

The lower bound on the critical value is easy.

Theorem 3.2.1. *Suppose* $\nu = \sum_k k(k-1)p_k/\mu < 1$. *Then the distribution of the size of the cluster containing 1 converges in distribution to a limit with mean* $1 + \mu/(1 - \nu)$.

Proof. Let A^n_0 be the set of neighbors of 1 and $S^n_k = |A^n_k|$. As long as $S^n_k > 0$ we pick an element $i^n_k \in A^n_k$, delete it from A^n_k and add its neighbors to A^n_k (some of which may already be in A^n_k) this set is called A^n_{k+1}. $S^n_{k+1} - S^n_k \le -1 + \xi^n_{k+1}$ where ξ^n_{k+1} is the number of uninspected neighbors of i^n_k. The mean of ξ^n_{k+1} is maximized when all of the k previously chosen elements added no new vertices to A. In this case we have removed k vertices of degree 1 from the distribution, so letting $\mathcal{F}^n_k$ be the σ-field generated by the first k choices, we have

$$E(\xi^n_{k+1}|\mathcal{F}^n_k) \le \bar{\nu}_n \frac{n}{n-k} \equiv \bar{\nu}^n_k$$

To be precise, in taking this expected value we are supposing that $d_1, \ldots, d_n$ and hence $\bar{\mu}_n, \bar{\nu}_n$ are fixed numbers and we are taking expectations for the random graph conditioned on their values.

It follows from the last inequality that $S^n_k - \sum_{j=1}^k (\bar{\nu}^n_j - 1)$ is a supermartingale. Letting $T_n = \inf\{k : S^n_k = 0\}$, stopping at $T_n \wedge m$, and noting that $\bar{\nu}^n_k$ is increasing in k, we have

$$ES^n_0 \ge E\left[S^n(T_n \wedge m) - (\bar{\nu}^n_m - 1)(T_n \wedge m) \right]$$

Since $S^n(T_n \wedge m) \ge 0$ we have $ES^n_0 \ge (1 - \bar{\nu}^n_m)E(T_n \wedge m)$ and it follows that

$$P(T_n > m) \le \frac{ES^n_0}{m(1 - \bar{\nu}^n_m)}$$

As $n \to \infty$, $ES_0^n \to \mu$ and for each fixed m we have $\bar{v}_m^n \to v < 1$,

$$\limsup_{n \to \infty} P(T_n > m) \le \frac{\mu}{m(1 - v)}$$

This shows that T_n is tight. Our stopping inequality implies

$$\limsup_{n \to \infty} E(T_n \wedge m) \le \frac{\mu}{(1 - v)}$$

so if T is any subsequential limit of the T_n, $ET \le \mu/(1 - v)$. It is easy to see that for any fixed m the distribution of ξ_m^n converges to q and the successive steps are independent, so the size of the cluster converges to the hitting time of 0 for the limiting random walk starting from S_0 with distribution p. This random variable has mean $\mu/(1 - v)$, so using Fatou's lemma now, we have

$$\liminf_{n \to \infty} E(T_n) \ge \frac{\mu}{(1 - v)}$$

and the proof of the result for the subcritical regime is complete. ∎

We come now to the more difficult proof of the upper bound on the critical value. The overall strategy of the proof is the same as the argument for Erdös–Rényi random graphs that was done in detail in Section 2.3. Because of this, we will content ourselves to explain how to cope with the new difficulties that come from a fixed degree distribution. Readers who want to see more details can consult the paper by Molloy and Reed (1995).

Theorem 3.2.2. *Suppose $\sum_k k(k - 1)p_k/\mu > 1$. Then there is a giant component of size $\sim (1 - G_0(\rho_1))n$, and no other clusters of size larger than $\beta \log n$.*

Sketch of proof. As in Section 2.3, we expose the cluster one site at a time to obtain something that can be approximated by a random walk. R_t is the set of removed sites, U_t are the unexplored sites and A_t is the set of active sites. $R_0 = \emptyset$, $U_0 = \{2, 3, \ldots, n\}$, and $A_0 = \{1\}$. At time $\tau = \inf\{t : A_t = \emptyset\}$ the process stops. If $A_t \ne \emptyset$, we pick i_t from A_t according to some rule that is measurable with respect to $\mathcal{A}_t$, we remove i_t from A_t, add its unexposed neighbors to A_t, and remove the added neighbors from U_t.

To estimate the change that occurs when some of the vertices have been exposed, let r_k be a distribution on the positive integers, and let $W_r(\omega)$ be a nondecreasing function of $\omega \in (0, 1)$ so that the Lebesgue measure $|\{\omega : W_r(\omega) = k\}| = r_k$. If we remove an amount of mass η from the distribution and renormalize to get a probability distribution then the result will be larger in distribution than $U = (W(\omega)|\omega < 1 - \eta)$.

From the last observation and the convergence of the size-biased empirical distribution to q in the total variation norm, we see that if n is large and the fraction of vertices exposed is at most η, then the cluster growth process dominates a random walk process in which steps (after the first one) have size -1 plus a random variable with distribution $q^\eta =_d (W_q | W_q < 1 - 2\eta)$.

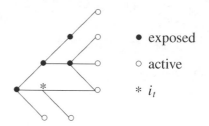

When we expose the neighbors of an active vertex, one of them might already be in the active set, as shown above. We call such an event a *collision*. If a collision occurs in the Erdös–Rényi growth process we are disappointed not to get another vertex. However, in the current situation, we must remove it from the active set, since the collision has reduced its degree to $d_x - 2$. To show that this does not slow down the branching process too much, we must bound the number of collisions. Note that q^η is concentrated on $\{0, \ldots, L\}$ where $L = W_q(1 - 2\eta)$. Thus until δn vertices have been exposed, the number of edges with an end in the active set is at most $\delta n L$. The probability of picking one of these edges in the exposure of an active vertex is at most $\delta n L / (t - \delta n L) \equiv \gamma$, where t is the total number edges. Let Z have distribution q^η with mean μ_η. The change in the set of active sites, correcting for collisions, is bounded by $X = -1 + Z - 2 \cdot \text{Binomial}(Z, \gamma)$. Therefore if $\delta < \eta$ is small, $EX = -1 + \nu_\eta(1 - 2\gamma) > 0$.

Define a random walk $S_t = S_0 + X_1 + \cdots + X_t$ where S_0 is the number of neighbors of the first site chosen and $X_1, \ldots, X_t$ are independent copies of our lower bounding variable, X. Since $EX > 0$, the random walk has positive probability of not hitting $(-\infty, 0]$, and there is positive probability that the cluster growth persists until there are at least δn vertices. To prove that we will get at least one such cluster with high probability, it is enough to show that each unsuccessful attempt will, with high probability, use up at most $O(\log n)$ vertices. For this guarantees that we will get a large number of independent trials before using a fraction δ of the vertices.

The random variable X is bounded so $\kappa(\theta) = Ee^{\theta X} < \infty$ for all θ. $\kappa(\theta)$ is convex, continuous and has $\kappa'(0) = EX > 0$, $\kappa(\theta) \geq P(X = -1)e^{-\theta} \to +\infty$ as $\theta \to -\infty$, so there is a unique $\lambda > 0$ so that $\kappa(-\lambda) = 1$. In this case $\exp(-\lambda S_k)$ is a nonnegative martingale. Let $T = \inf\{k : S_k \leq 0\}$. Due to the possible removal of active vertices, the random walk may jump down by more than 1, but its jumps are bounded, so $\exp(-\lambda S_{k \wedge T})$ is bounded, and the optional stopping theorem implies that the probability of reaching ≤ 0 from $S_0 = x$ is $\leq e^{-\lambda x}$.

The last estimate implies that the probability that the set of active vertices grows to size $(2/\lambda)\log n$ without generating a large cluster is $\leq n^{-2}$. Routine large deviation estimates for sums of independent random variables show that if β is large, the probability that the sum of $\beta \log n$ independent copies of X is $\leq (2/\lambda) \log n$ is at most n^{-2}, see Lemma 2.3.3 and Step 2 in the proof of Theorem 2.3.2. Thus the probability of exposing more than $\beta \log n$ vertices and not generating a large cluster is $\leq 2n^{-2}$.

At this point we can finish up as we did in Section 2.3. Two clusters that reach size $\beta \log n$ will with high probability grow to size $n^{2/3}$ and hence intersect with probability $1 - o(n^{-2})$, see Step 3. The results above show that size of the giant component is with high probability the same as $\{x : |\mathcal{C}_x| \geq \beta \log n\}$. As in Step 4, the growth of the cluster up to size $\beta \log n$ can be approximated by the two-phase branching process, and these events are almost independent, so computing second moments gives the final conclusion of the theorem. ∎

What Did Molloy and Reed Do? They used a different model in which one specifies $v_i(n)$, the number of vertices of degree i. They assumed

  (i) $v_i(n) \geq 0$
 (ii) $\sum_i v_i(n) = n$
(iii) the degree sequence is *feasible*, that is, $\sum_i i v_i(n)$ is even
(iv) the degree sequence is *smooth*, that is, $\lim_{n\to\infty} v_i(n)/n = p_i$
 (v) the degree sequence is *sparse*, that is, $\sum_{i\geq 0} i v_i(n)/n \to \sum_i i p_i$.

Our random degree model has these properties with probability one. However, Molloy and Reed's approach is more refined since it deals with individual sequence of degree sequences. Unfortunately, it also needs more conditions on maximum degrees.

To state their main result (Theorem 1 on pages 164–165) recall that our condition for criticality is $1 = \sum (k-1)k p_k / \mu$. Multiplying each side by μ, recalling that $\mu = \sum_k k p_k$, then subtracting gives

$$Q \equiv \sum_k k(k-2) p_k = 0$$

The final detail is that they use asymptotically almost surely (a.a.s) for "with probability tending to 1 as $n \to \infty$."

Theorem 3.2.3. *Suppose (i)–(v) hold, and that $v_i(n) = 0$ for $i > n^{1/4-\epsilon}$.*

(a) *If $Q > 0$ there are constants $\zeta_1, \zeta_2 > 0$ so that a.a.s. G has a component with at least $\zeta_1 n$ vertices and $\zeta_2 n$ cycles.*

(b) *Suppose $Q < 0$ and $v_i(n) = 0$ for $i \geq w_n$ where $w_n \leq n^{1/8-\epsilon}$. Then there is a $\beta > 0$ so that a.a.s. there is no cluster with $\beta w_n^2 \log n$ vertices.*

Molloy and Reed (1995) did not get the exact size of the giant component since they used their lower bound on the growth starting from one point until the cluster reached size δn. In their 1998 article they found the exact size of the giant component and proved a "duality result" for small clusters in the supercritical regime. We will state the result and then sketch its proof. Let $\mu = \sum_i i p_i$ and define

$$\chi(\alpha) = \mu - 2\alpha - \sum_{i \geq 1} i p_i \left(1 - \frac{2\alpha}{\mu}\right)^{i/2}$$

Theorem 3.2.4. *Let α^* be the smallest positive solution of $\chi(\alpha) = 0$ in $[0, \mu/2]$. The size of the giant component is $\epsilon^* n + o(n)$ where*

$$\epsilon^* = 1 - \sum_{i \geq 1} p_i \left(1 - \frac{2\alpha^*}{\mu}\right)^{i/2}$$

In the supercritical regime the finite clusters have the same distribution as a random graph with $(1 - \epsilon^*)n$ vertices and degree distribution

$$p_i' = \frac{p_i}{1 - \epsilon^*} \left(1 - \frac{2\alpha^*}{\mu}\right)^{i/2}$$

Comparison with Heuristics. At first these formulas do not look much like the ones in Section 3.1. However, rewriting $\chi(\alpha^*) = 0$

$$\left(1 - \frac{2\alpha^*}{\mu}\right)^{1/2} = \sum_{i \geq 1} \frac{i p_i}{\mu} \left(1 - \frac{2\alpha^*}{\mu}\right)^{(i-1)/2}$$

we see that $\xi = (1 - 2\alpha^*/\mu)^{1/2}$ is a fixed point of G_1 the generating function of $q_{k-1} = k p_k/\mu$, and the expression for ϵ^* is just $1 - G_0(\xi)$. For the third and final equation, note that the left-hand side gives the distribution of the first generation in the two-phase branching process, when it is conditioned to die out.

Sketch of proof. Molloy and Reed (1998) expose the clusters one edge at a time. To begin they form a set L consisting of i distinct copies of each of the $v_i(n)$ vertices with degree i. In the construction that follows, the pairing of the vertices that define the graph are being done as we proceed. At each step, a vertex, all of whose copies are in exposed pairs, is *entirely exposed*. A vertex, with some but not all of its copies in exposed pairs, is *partially exposed*. The copies of partially exposed vertices which are not in exposed pairs are *open*. All other vertices are *unexposed*.

Repeat until L is empty.

- Expose a pair of vertices by first choosing any member of L, and then choosing its partner at random. Remove them from L.

- Repeat until there are no partially exposed vertices: choose an open copy of a partially exposed vertex, and pair it with another randomly chosen member of L. Remove them both from L.

Let $M = \sum_i i v_i(n)$. Let X_j be the number of open vertices and let $v_{i,j}$ be the number of unexposed vertices of degree i when j pairs have been exposed. Let $\mathcal{F}_j$ be the σ field generated by the first j choices. If $X_j > 0$ then

$$E(v_{i,j+1} - v_{i,j}|\mathcal{F}_j) = -\frac{i v_{i,j}}{M - 2j - 1} \qquad (3.2.1)$$

The denominator gives the number of unused vertex copies when the choice at time $j + 1$ is made. The relevance of $X_j > 0$ is that it guarantees that the first vertex chosen is partially exposed. When $X_j = 0$ we have two opportunities to lose an unexposed vertex of degree i. To get around this problem, Molloy and Reed show that for any function $\omega(n) \to \infty$ a.a.s. the giant component will be one of the first $\omega(n)$ components exposed. Since all nongiant components have size $O(\log n)$, it follows that a.a.s. the $(\log^2 n)$th edge will lie in the giant component. Because of this we can watch the process from time $(\log^2 n)$ until the next time $X_j = 0$ and not worry about the second case.

As $n \to \infty$, $M/n \to \mu$. Thus (3.2.1) tells us that when $j = \alpha n$, the infinitesimal mean change in $v_{i,j}/n$ converges to

$$\frac{i}{\mu - 2\alpha} \cdot \frac{v_{i,j}}{n}$$

The infinitesimal variance of $v_{i,j}/n$ is $\leq 1/4n^2$ so as in Section 2.5, we can conclude that the rescaled process has a deterministic limit. Readers who want more details about this step can consult Theorem 1 in Wormald (1995).

Writing $Z_i(\alpha) = \lim_{n\to\infty} v_{i,n\alpha}/n$, we have

$$Z_i'(\alpha) = -\frac{i Z_i(\alpha)}{\mu - 2\alpha}$$

The solution with initial condition $Z_i(0) = p_i$ is

$$Z_i(\alpha) = p_i \left(1 - \frac{2\alpha}{\mu}\right)^{i/2}$$

To check this note

$$Z_i'(\alpha) = p_i \frac{i}{2} \left(1 - \frac{2\alpha}{\mu}\right)^{(i/2)-1} \left(-\frac{2}{\mu}\right) = -Z_i(\alpha) \cdot \frac{i}{2} \cdot \frac{\mu}{\mu - 2\alpha} \cdot \frac{2}{\mu}$$

It is clear from the definitions that $X_j = M - 2j - \sum_i i v_{i,j}$. Taking limits

$$X_{\alpha n}/n \approx \mu - 2\alpha - \sum_i i p_i \left(1 - \frac{2\alpha}{\mu}\right)^{i/2}$$

To justify the interchange of limit and the infinite sum, note that $v_{i,j}$ is decreasing in j, and if we pick I large enough $(1/n) \sum_{i=I}^{\infty} v_{i,0} \leq \epsilon$ and hence $(1/n) \sum_{i=I}^{\infty} v_{i,j} \leq \epsilon$ for all j.

Recall that all of our computations have been under the assumption $X_j > 0$, $j \leq \alpha n$. The right-hand side hits zero at $\alpha = \alpha^*$ signaling the end of the exposure of the giant component. When we are done exposing the giant component the number of vertices of degree i that remain is

$$nZ_i(\alpha^*) = np_i \left(1 - \frac{2\alpha^*}{\mu} \right)^{i/2} \equiv np_i'$$

and the total number is $n(1 - \epsilon^*) = \sum_i p_i'$.   ∎

3.3 Subcritical Estimates

In subcritical Erdös–Rényi random graphs the largest component is $O(\log n)$. This result is false for graphs with arbitrary degree distributions. Suppose for example that $p_k \sim ck^{-\gamma}$ where $\gamma > 3$ so that the variance is finite. The tail of the distribution $\sum_{k=K}^{\infty} p_k \sim cK^{1-\gamma}$ so the largest degree present in a graph with n vertices is $O(n^{1/(\gamma-1)})$. At first it might seem that one can increase the power of n by looking at the size biased distribution $q_{k-1} = kp_k/\mu$ which is $\sim ck^{1-\gamma}$. However, the degrees are i.i.d. with distribution p_k and the size biasing can only change the probabilities of degrees that are present. Because of this we conjecture:

Conjecture 3.3.1. *If $p_k \sim ck^{-\gamma}$ with $\gamma > 3$ and $\nu < 1$ then the largest component is $O(n^{1/(\gamma-1)})$.*

Chung and Lu have introduced a variant of the Molloy and Reed model that is easier to study. Their model is specified by a collection of weights $\mathbf{w} = (w_1, \ldots, w_n)$ that represent the expected degree sequence. The probability of an edge between i to j is $w_i w_j / \sum_k w_k$. They allow loops from i to i so that the expected degree at i is

$$\sum_j \frac{w_i w_j}{\sum_k w_k} = w_i$$

Of course for this to make sense we need $(\max_i w_i)^2 < \sum_k w_k$.

Let $d = (1/n) \sum_k w_k$ be the average degree. As in the Molloy and Reed model, if we follow an edge from i, vertices are chosen proportional to their weights, that is, j is chosen with probability $w_j / \sum_k w_k$. Thus the relevant quantity for connectedness of the graph is the second-order average degree

$$\bar{d} = \sum_j w_j \frac{w_j}{\sum_k w_k}$$

The Cauchy–Schwarz inequality implies

$$\left(\sum_k w_k\right)^2 \leq \left(\sum_k w_k^2\right)\left(\sum_k 1\right)$$

so $\bar{d} \geq d$. It is not clear whether Chung and Lu (2002a) understood that $\bar{d} > 1$ is the correct condition for the existence of a giant component, since they prove $d > 1$ is sufficient and derive a number of bounds on the size of the giant component under this condition.

Chung and Lu (2002a) have proved a nice result about the subcritical phase. Note that when $\gamma > 3$, $1/(\gamma - 1) < 1/2$ so this result is consistent with our conjecture.

Theorem 3.3.2. *Let* $vol(S) = \sum_{i \in S} w_i$. *If* $\bar{d} < 1$ *then all components have volume at most* $C\sqrt{n}$ *with probability at least*

$$1 - \frac{d\bar{d}^2}{C^2(1 - \bar{d})}$$

Proof. Let x be the probability that there is a component with volume $> C\sqrt{n}$. Pick two vertices at random with probabilities proportional to their weights. If there is a "large component" with volume $\geq C\sqrt{n}$, then for each vertex, the probability it is in a large component is $\geq C\sqrt{n}\gamma$, where $\gamma = 1/\sum_i w_i$. Therefore the probability a randomly chosen pair of vertices is in the same component is at least

$$x(C\sqrt{n}\gamma)^2 = C^2 x n \gamma^2 \qquad (3.3.1)$$

On the other hand, for a fixed pair of vertices u and v the probability $p_k(u, v)$ of u and v being connected by a path of length $k + 1$ is a most

$$p_k(u, v) \leq \sum_{i_1, i_2, \ldots, i_k} (w_u w_{i_1} \gamma)(w_{i_1} w_{i_2} \gamma) \cdots (w_{i_k} w_v \gamma) \leq w_u w_v \gamma (\bar{d})^k$$

Summing over $k \geq 0$ the probability u and v belong to the same component is at most

$$\frac{w_u w_v \gamma}{1 - \bar{d}}$$

The probabilities of u and v being selected are $w_u \gamma$ and $w_v \gamma$. Summing over u and v the probability a randomly chosen pair of vertices belong to the same component is at most

$$\frac{(\bar{d})^2 \gamma}{1 - \bar{d}}$$

Using this with (3.3.1)

$$C^2 x n \gamma^2 \leq \frac{(\bar{d})^2 \gamma}{1 - \bar{d}}$$

which implies

$$x \leq \frac{(\bar{d})^2}{C^2 n (1 - \bar{d}) \gamma} = \frac{d(\bar{d})^2}{C^2 (1 - \bar{d})}$$

since $\gamma = 1/\sum_i w_i$ and $d = (1/n)\sum_i w_i$ implies $n\gamma = 1/d$.   ∎

3.4 Distances: Finite Variance

In this section we will consider the typical distance between vertices in the giant component of a Newman–Strogatz–Watts random graph when the random degree D has $P(D \geq x) \leq x^{-\beta+1}$ with $\beta > 3$, so that the distribution has finite variance. The results here follow van der Hofstad, Hooghiemstra and Van Mieghem (2005). Let $\mu = ED$ and let $\nu = E(D(D-1))/ED$ be the mean of the size biased distribution. Let Z_t be the two-phase branching process and let

$$W = \lim_{t \to \infty} \frac{Z_t}{\mu \nu^{t-1}} \tag{3.4.1}$$

Theorem 3.4.1. *Suppose $\nu > 1$. Let H_n be the distance between 1 and 2 in the random graph on n vertices. For $k \geq 1$, let $a(k) = [\log_\nu k] - \log_\nu \in (-1, 0]$. As $n \to \infty$*

$$P(H_n - [\log_\nu n] = k | H_n < \infty) = P(R_{a(n)} = k) + o(1)$$

If $\kappa = \mu(\nu - 1)^{-1}$ then for $a \in (-1, 0]$

$$P(R_a > k) = E(\exp(-\kappa \nu^{a+k} W_1 W_2) | W_1 W_2 > 0)$$

where W_1 and W_2 are independent copies of W defined in (3.4.1).

$x = \log_\nu k$ has $\nu^x = k$, that is, $x = \log k / \log \nu$, so the answer has the same intuitive explanation as the one in Theorem 2.4.1 for Erdős–Rényi random graph. The strength of this result is that it shows that the fluctuations around the limit are $O(1)$. The statement is made complicated by the fact that H_n is always an integer so the support of $H_n - \log_\nu n$ changes with n.

The proof of Theorem 3.4.1 is done by a careful comparison of the growing cluster and branching process. The proof is about 40 pages, so we will content ourselves to explain where the formula for the distribution of R_a comes from and refer the reader to the original paper for details. The total number of vertices is $\sim \mu n$ as $n \to \infty$, so for our sketch we will suppose that it is always equal to μn. Taking turns growing each cluster by one branching step,

$$P(H_n > k) \approx E \exp\left(-\sum_{i=2}^{k+1} \frac{Z^1_{\lceil i/2 \rceil} Z^2_{\lfloor i/2 \rfloor}}{\mu n}\right)$$

Here, one needs to look closely at the indices. We round up in the first case, and down in the second, so for $i = 2, 3, 4, \ldots$ we get $(\lceil i/2 \rceil, \lfloor i/2 \rfloor) = (1, 1), (2, 1), (2, 2) \ldots$. When $k = 1$ there is only one term in the sum: $Z^1_1 Z^2_1 / \mu n$. This gives the expected number of the Z^1_1 half-edges from 1 that are paired with one of the Z^2_1 half-edges from 2. The distance is > 1 if and only if the number of such pairings is 0. Since the events are almost independent, the number of pairings (conditional on the values of Z^1_1 and Z^2_1) is roughly Poisson and the probability of 0 is $E \exp(-Z^1_1 Z^2_1 / \mu n)$.

Letting $\sigma_n = [\log_v n]$ and writing $n = v^{\log_v n} = v^{\sigma_n - a_n}$ we have

$$P(H_n > \sigma_n + j) \approx E \exp\left(-\mu v^{a_n + k} \sum_{i=2}^{\sigma_n + j + 1} \frac{Z^1_{\lceil i/2 \rceil} Z^2_{\lfloor i/2 \rfloor}}{\mu^2 v^{\sigma_n + j}}\right)$$

$\lceil i/2 \rceil + \lfloor i/2 \rfloor = i$ so using (3.4.1)

$$\frac{Z^1_{\lceil (\sigma_n + j + 1)/2 \rceil} Z^2_{\lfloor (\sigma_n + j + 1)/2 \rfloor}}{\mu^2 v^{\sigma_n + j - 1}} \to W_1 W_2$$

and it follows that

$$\sum_{i=2}^{\sigma_n + j + 1} \frac{Z^1_{\lceil i/2 \rceil} Z^2_{\lfloor i/2 \rfloor}}{\mu^2 v^{\sigma_n + j}} \to v^{-1} W_1 W_2 \sum_{\ell=0}^{\infty} (1/v)^\ell = \frac{1}{v - 1}$$

Combining our results gives

$$P(H_n > \sigma_n + j \,|\, H_n < \infty) \approx E(\exp(-\kappa v^{a_n + j} W_1 W_2) \,|\, W_1 W_2 > 0)$$

with $\kappa = \mu/(v - 1)$.

3.5 Epidemics

In this section we will consider the SIR (susceptible–infected–removed) epidemic model on Newman–Strogatz–Watts random graphs, following Newman (2002). The closely related topic of percolation on these random graphs was studied earlier by Callaway, Newman, Strogatz, and Watts (2000). An infected individual stay infected for a random amount of time τ, and during this time infects susceptible neighbors at rate r. At the end of the infected period the individual becomes removed, that is, no longer susceptible to the disease. To have a happier story we will not think of a fatal disease but instead consider a disease like measles or a particular strain of influenza, where upon recovery one has immunity to further infection.

The key to our study of the SIR model is that if we look at the neighbors that a given individual will try to infect, we will obtain a new NSW random graph. Once this is established, we can immediately compute the threshold, the fraction of individuals who will become infected, etc. We begin with the case in which the infection time τ is constant, and without loss of generality scale time to make the constant 1. Let p_k be the degree distribution. If we start with one individual infected then the probability that j of its k neighbors will become infected is

$$\hat{p}_j = \sum_{k=j}^{\infty} p_k \binom{k}{j} (1 - e^{-r})^j \, e^{-(k-j)r}$$

If μ is the mean of p then the mean of $\hat{p}$ is $\hat{\mu} = \mu(1 - e^{-r})$.

Due to the construction of the social network as an NSW random graph, individuals in the first and subsequent generations will have k neighbors (including their parent) with probability $q_k = (k+1)p_{k+1}/\mu$ for $k \geq 0$. The probability that j of their neighbors will become infected is

$$\hat{q}_j = \sum_{k=j}^{\infty} q_k \binom{k}{j} (1 - e^{-r})^j \, e^{-(k-j)r}$$

If ν is the mean of q then the mean of $\hat{q}$ is $\hat{\nu} = \nu(1 - e^{-r})$, and the condition for the disease to propagate is

$$\nu(1 - e^{-r}) > 1 \tag{3.5.1}$$

Newman writes things in terms of the transmission probability T, here $1 - e^{-r}$, so this agrees with his equation (23).

One can compute the probability that an epidemic occurs in the usual way using the generating functions of $\hat{q}$ and $\hat{p}$. Writing $T = 1 - e^{-r}$

$$\hat{G}_0(z) = \sum_{j=0}^{\infty} \sum_{k=j}^{\infty} p_k \binom{k}{j} T^j (1 - T)^{k-j} z^j$$
$$= \sum_{k=0}^{\infty} p_k \sum_{j=0}^{k} \binom{k}{j} (Tz)^j (1 - T)^{k-j} = G_0(Tz + (1 - T))$$

An almost identical calculation gives $\hat{G}_1(z) = G_1(Tz + (1 - T))$.

Theorem 3.5.1. *If $\hat{G}_1(\xi) = \xi$ is the smallest fixed point in $[0, 1]$ the probability the epidemic does not die out is $1 - \hat{G}_0(\xi)$.*

We turn now to the case of random τ. The probability that an infected individual infects a given neighbor is

$$T = 1 - \int_0^\infty dt\, P(\tau = t)e^{-rt}$$

Again T is for transmissibility. In the discussion leading to (13), Newman claims that the infection of different neighbors are independent events, but it is easy to see that this is false. Suppose, for example, that the infection time is exponential with mean one, $P(\tau = t) = e^{-t}$. In this case

$$1 - T = \int_0^\infty dt\, e^{-t}e^{-rt} = \frac{1}{r+1}$$

while the probability two neighbors both escape infection is

$$\int_0^\infty dt\, e^{-t}e^{-2rt} = \frac{1}{2r+1} > \frac{1}{r^2+2r+1} = \left(\frac{1}{r+1}\right)^2$$

In general if τ is not constant, the events that neighbors x and y escape infection are positively correlated since Jensen's inequality implies

$$E(e^{-2r\tau}) > (Ee^{-r\tau})^2$$

To compute the correct distribution of the number of neighbors that will become infected, write r_k in place of p_k or q_k. The generating function of the number of neighbors that will become infected is

$$\hat{G}(z) = \int_0^\infty dt\, P(\tau = t) \sum_{j=0}^\infty z^j \sum_{k=j}^\infty r_k \binom{k}{j}(1 - e^{-rt})^j\, e^{-r(k-j)t} \qquad (3.5.2)$$

Interchanging the order of summation, putting z^j with $(1 - e^{-rt})^j$, and using the binomial theorem we have

$$= \int_0^\infty dt\, P(\tau = t) \sum_{k=0}^\infty r_k(e^{-rt} + z(1 - e^{-rt}))^k$$

$$= \int_0^\infty dt\, P(\tau = t)G(e^{-rt} + z(1 - e^{-rt}))$$

$$= EG(e^{-r\tau} + z(1 - e^{-r\tau})) = EG(1 + (z - 1)(1 - e^{-r\tau}))$$

If $r_0 + r_1 < 1$, G is strictly convex, so recalling $E(1 - e^{-r\tau}) = T$, the above is

$$> G(1 + (z - 1)T) \equiv \tilde{G}(z)$$

the result in Newman's formulas (13) and (14).

Although Newman's formula for the generating function is not correct the threshold is, since $\hat{v} = vT$. Let $\hat{G}_1$ and $\hat{G}_0$ for the generating functions of $\hat{q}$ and $\hat{p}$ and let

$$\tilde{G}_i = G_i(1 + (z-1)T)$$

be Newman's formulas. The extinction probability $\hat{\xi}$ for $\hat{G}_1$ is larger than that for $\tilde{G}_1$, call it $\tilde{\xi}$, and hence $\hat{G}_0(\hat{\xi}) > \tilde{G}_0(\tilde{\xi})$. Given this inequality it is remarkable that the survival probabilities for Newman's simulations in his Figure 1 match his theoretical predictions almost exactly. The next example makes this somewhat less surprising.

Concrete Example. To investigate the differences between $\hat{G}$ and $\tilde{G}$, we will consider a concrete example. Suppose $p_k = 1/3$ for $k = 1, 2, 3$. In this case $\mu = 2$, $q_0 = 1/6$, $q_1 = 2/6$ and $q_2 = 3/6$, so $v = 4/3$. For simplicity, we will take a rather extreme infection time distribution $P(\tau = \infty) = p$ and $P(\tau = 0) = 1 - p$. In this case, the epidemic process is site percolation, while Newman's version is bond percolation. Both versions have critical values $p_c = 3/4$, but as we will now compute have somewhat different percolation probabilities.

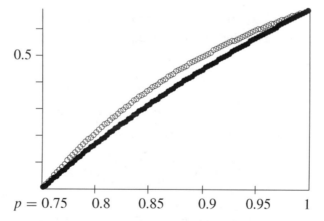

Comparison of percolation probabilities for site (circles) and bond (black dots) percolation.

Using (3.5.2)

$$\hat{G}_1(z) = 1 - p + p\left(\frac{1}{6} + \frac{z}{3} + \frac{z^2}{2}\right)$$

compared with

$$\tilde{G}_1(z) = G_1(1 - p + pz) = \frac{1}{6} + \frac{1 - p + pz}{3} + \frac{(1 - p + pz)^2}{2}$$

$$= \left(\frac{1}{6} + \frac{1 - p}{3} + \frac{(1 - p)^2}{2}\right) + \frac{1}{3}pz + (1 - p)pz + \frac{1}{2}p^2z^2$$

The generating functions are quadratic $az^2 + bz + c$ and have 1 as a fixed point, so $a + b + c = 1$ and the fixed point equation $az^2 + (b - 1)z + c = 0$ can be factored

$$(z - 1) \cdot (az + b + a - 1) = 0$$

to conclude that the fixed point of interest is $z = (1 - a - b)/a$. From this we conclude

$$\hat{\xi} = \frac{1 - 5p/6}{p/2} \qquad \tilde{\xi} = \frac{1 - p^2/2 - (1 - p)p - p/3}{p^2/2}$$

These formulas do not look much alike but when we plot $1 - \hat{\xi}$ (black dots) and $1 - \tilde{\xi}$ (circles) the numerical values are similar.

4

Power Laws

4.1 Barabási–Albert Model

As we have noted, many real-world graphs have power law degree distributions. Barabási and Albert (1999) introduced a simple model that produces such graphs. They start with a graph with

a small number of vertices m_0. At every time step, we add a new vertex with m edges that link the new vertex to m different vertices already present in the system. To incorporate preferential attachment, we assume that the probability Π that a new vertex will be connected to a vertex i depends on the connectivity of that vertex, so that $\Pi(k_i) = k_i / \sum_j k_j$. After t steps the model leads to a random network with $t + m_0$ vertices and mt edges.

Bollobás, Riordan, Spencer, and Tusnády (2001) complain: "The description of the random graph process quoted above is rather imprecise. First as the degrees are initially zero, it is not clear how the process is supposed to get started. More seriously, the expected number of edges linking a new vertex to earlier vertices is $\sum_i \Pi(k_i) = 1$, rather than m. Also when choosing in one go a set S of m earlier vertices as neighbors of v, the distribution is *not* specified by giving the marginal probability that each vertex lies in S."

As we will see below there are several ways to make the process precise and all of them lead to the same asymptotic behavior. For the moment, we will suppose that the process starts at time 1 with two vertices linked by m parallel edges, so that the total degree at any time t is $2mt$. When we add a new vertex we will add edges one a time, with the second and subsequent edges doing preferential attachment using the updated degrees. This scheme has the desirable property that a graph of size n for a general m can be obtained by running the $m = 1$ model for nm steps and then collapsing vertices $km, km - 1, \ldots (k - 1)m + 1$ to make vertex k.

Having settled on a definition we turn now to a discussion of the results. Barabási and Albert (1999) did simulations that suggested a power law distribution with power $\gamma = 2.9 \pm 0.1$ and gave the following argument for $\gamma = 3$. If we consider the degree of i, k_i, to be a continuous variable then

$$\frac{\partial k_i}{\partial t} = m \cdot \frac{k_i}{\sum_j k_j} = \frac{k_i}{2t}$$

since $\sum_j k_j = 2mt$. The solution to this differential equation is

$$k_i(t) = m(t/t_i)^{1/2}$$

where t_i is the time the vertex was introduced. From this we see that

$$P(k_i(t) > k) = P(t_i < tm^2/k^2) = m^2/k^2 \qquad (4.1.1)$$

since vertices are added uniformly on $[0, t]$. Differentiating

$$P(k_i(t) = k) = 2m^2/k^3$$

Dorogovtsev, Mendes, and Samukhin (2000) took a different approach, using what they call the "master equation," which can be used to obtain rigorous asymptotics for the mean number of vertices of degree k, $N(k, t)$. By considering what happens on one step, in which vertices of degree k can be created from those of degree $k - 1$ or lost by becoming degree $k + 1$, we see that

$$N(k, t+1) - N(k, t) = \frac{m(k-1)}{2mt} N(k-1, t) - \frac{mk}{2mt} N(k, t) + \delta_{k,m}$$

Here we ignore the possibility of more than one edge being attached to one vertex, and the updating of edges that occurs as the m edges are added. The last term takes care of the fact that when $k = m$ we add one vertex of degree m at each time. To make the equation correct for $k = m$ we suppose $N(j, t) = 0$ for $j < m$.

We will encounter equations like this several times below, so we will generalize to prepare for future examples. When $k = m$, we can write the equation as

$$N(m, t+1) = c + \left(1 - \frac{b(t)}{t}\right) N(m, t) \qquad (4.1.2)$$

where $b(t) \to b$ as $t \to \infty$. In the current case, $c = 1$ and $b(t) \equiv m/2$.

Lemma 4.1.1. $N(m, t)/t \to c/(1 + b)$.

Proof. We begin with case in which $b(t) \equiv b$ is constant. Iterating our equation once gives

$$N(m, t+1) = c + \left(1 - \frac{b}{t}\right) c + \left(1 - \frac{b}{t}\right)\left(1 - \frac{b}{t-1}\right) N(m, t-1)$$

and repeating we have

$$N(m, t+1) = c \sum_{s=1}^{t} \prod_{r=s+1}^{t} \left(1 - \frac{b}{r}\right) + N(m, 1) \prod_{r-1}^{t} \left(1 - \frac{b}{r}\right)$$

The product is $\approx \exp(-\sum_{r=s+1}^{t} b/r) \approx \exp(-b(\log t - \log s)) = (s/t)^b$ so

$$N(m, t) \sim ct^{-b} \int_0^t s^b \, ds = ct^{-b} \frac{1}{b+1} t^{b+1} = ct/(1+b)$$

which proves the desired result for constant $b(t)$. To extend to the case in which $b(t) \to b$, note that the solution decreases when $b(t)$ increases, so comparing with functions $\bar{b}(t)$ that are constant for large t, and using the fact that the answer is a continuous function of the limit b gives the result in general. ∎

We abstract the equation for $k > m$ as

$$N(k, t+1) = \left(1 - \frac{b(t)}{t}\right) N(k, t) + g(t)$$

where $b(t) \to b$ and $g(t) \to g$ as $t \to \infty$. In the current case $b(t) \equiv k/2$ and $g(t) = (k-1)N(k-1, t)/2t$ which has a limit by induction.

Lemma 4.1.2. $N(k, t)/t \to g/(b+1)$.

Proof. By remarks at the end of the previous proof, it suffices to prove the result when $b(t)$ is constant. Iterating (4.1.2) we have

$$N(k, t+1) = g(t) + \left(1 - \frac{b}{t}\right) g(t-1) + \left(1 - \frac{b}{t}\right)\left(1 - \frac{b}{t-1}\right) N(k, t-1)$$

Using $N(k, 1) = 0$ for $k > m$ leads to

$$N(k, t+1) = \sum_{i=0}^{t-1} g(t-i) \prod_{j=0}^{i-1} \left(1 - \frac{b}{t-j}\right)$$

Changing variables $s = t - i, r = t - j$ the above is

$$N(k, t+1) = \sum_{s=1}^{t} g(s) \prod_{r=s+1}^{t} \left(1 - \frac{b}{r}\right)$$

Again the product is $\sim (s/t)^b$, so

$$N(k, t+1)/t \sim t^{-(b+1)} \int_0^t g(s)s^b \, ds \to g/(b+1)$$

proving the result. ∎

Returning to the current situation and letting $p_k = \lim_{t\to\infty} N(k,t)/t$, Lemma 4.1.1 implies that

$$p_m = 2/(m+2)$$

while Lemma 4.1.2 tells us that for $k > m$

$$p_k = \frac{(k-1)p_{k-1}}{2} \cdot \frac{2}{k+2} = p_{k-1}\frac{k-1}{k+2}$$

The solution to this recursion is

$$p_k = \frac{2m(m+1)}{k(k+1)(k+2)} \tag{4.1.3}$$

To check this, note that it works when $k = m$ and if the formula is correct for $k-1$ then it is correct for k.

With the asymptotics for the mean in hand, the rest is easy thanks to the inequality of Azuma (1967) and Hoeffding (1963).

Lemma 4.1.3. *Let X_t be a martingale with $|X_s - X_{s-1}| \le c$ for $1 \le s \le t$. Then*

$$P(|X_t - X_0| > x) \le \exp(-x^2/2c^2 t)$$

Let $Z(k,t)$ be the number of vertices of degree k at time t, and let $\mathcal{F}_s$ denote the σ-field generated by the choices up to time s. We apply the result to $X_s = E(Z(k,t)|\mathcal{F}_s)$. We claim that $|X_s - X_{s-1}| \le 2m$. To see this, first consider $m = 1$ and note that whether we attach the vertex v_s added at time s to v or v' does not effect the degrees of $w \ne v, v'$, or the probabilities they will be chosen later, so it follows that $|X_s - X_{s-1}| \le 2$. The results for general m follows from its relationship to model with $m = 1$.

Since $Z(k,0) = E(Z(k,t))$ taking $x = \sqrt{t \log t}$ we have

$$P(|Z(k,t) - E(Z(k,t))| > \sqrt{t \log t}) \le t^{-1/8}$$

and hence

Theorem 4.1.4. *As $t \to \infty$, $Z(k,t) \to p_k$ in probability.*

We learned this simple proof from Bollobás, Riordan, Spencer, and Tusnády (2001).

4.2 Related Models

In the previous section we saw that the preferential attachment model of Barabási and Albert produced a degree distribution $p_k \sim Ck^{-\beta}$ with $\beta = 3$. However, as we saw in Section 1.4, there are many examples of powers between 2 and 3. Krapivsky, Redner, and Leyvraz (2000) and Krapivsky and Redner (2001) approached this problem by modifying the rules so that attachment to a degree k vertex is proportional to $f(k)$, for example, $f(k) = k^\gamma$. Here, we will use the fact that graph with n vertices and m edges are added at once, $G_n^{(m)}$ can be obtained by collapsing vertices in $G_{nm}^{(1)}$ to reduce to the case in which one vertex and one edge are added at each step, and G_1 is one edge connecting two vertices.

Let $N_k(t)$ be the expected number of vertices of degree k at time t. If we consider the model with $f(k) = k^\gamma$ and let $M_f(t) = \sum_k f(k)N_k(t)$ then

$$N_k(t+1) - N_k(t) = \frac{1}{M_f(t)}[f(k-1)N_{k-1}(t) - f(k)N_k(t)] + \delta_{k,1} \qquad (4.2.1)$$

where $\delta_{k,1} = 1$ if $k = 1$ and 0 otherwise, and we set $N_0(t) \equiv 0$.

Case 1. $0 < \gamma < 1$. When $f(x) = x^\gamma$ we write M_γ for M_f. It is easy to see that $M_0(t) = t + 1$, $M_1(t) = 2t$, and $M_0(t) \le M_\gamma(t) \le M_1(t)$. With $M_\gamma(t)/t \in [1, 2]$ for all t it is reasonable to guess that

(A) As $t \to \infty$, $M_\gamma(t)/t \to \mu \in (1, 2)$.

Only a mathematician would be pessimistic enough to think that the bounded sequence $M_\gamma(t)/t$ might not converge, so following the three physicists, we will assume (A) for the rest of our calculations. Let $n_k = \lim_{t\to\infty} N_k(t)/t$. Writing

$$N_1(t+1) = 1 + \left(1 - \frac{f(1)}{M_f(t)}\right) N_1(t)$$

and using (A) with Lemma 4.1.1 and noting $c = 1$, $b = f(1)/\mu$ gives

$$n_1 = \frac{c}{1+b} = \frac{\mu}{\mu + f(1)} \qquad (4.2.2)$$

When $k > 1$, (4.2.1) gives

$$N_k(t+1) = \left(1 - \frac{f(k)}{M_f(t)}\right) N_k(t) + \frac{f(k-1)}{M_f(t)}N_{k-1}(t)$$

Using Lemma 4.1.2 and noting $g = n_{k-1}f(k-1)/\mu$ and $b = f(k)/\mu$ gives

$$n_k = \frac{g}{1+b} = \frac{f(k-1)}{\mu + f(k)}n_{k-1}$$

Iterating, then shifting the indexing of the numerator, recalling $f(1) = 1$ and $n_1 = \mu/(\mu + f(1))$, we have

$$n_k = \prod_{j=2}^k \frac{f(j-1)}{\mu + f(j)} n_1 = \frac{\mu}{f(k)} \prod_{j=1}^k \frac{f(j)}{\mu + f(j)} = \frac{\mu}{k^\gamma} \prod_{j=1}^k \left(1 + \frac{\mu}{j^\gamma}\right)^{-1} \qquad (4.2.3)$$

To determine the asymptotic behavior of n_k we note that since $\gamma < 1$

$$\log \prod_{j=1}^{k}(1 + \mu/j^\gamma)^{-1} \sim -\sum_{j=1}^{k} \mu/j^\gamma \sim -\frac{\mu}{1-\gamma}k^{(1-\gamma)}$$

Putting the pieces together we see that the limiting degree distribution is, for large k,

$$\approx \frac{\mu}{k^\gamma} \exp(-ck^{(1-\gamma)})$$

which decays faster than any power but not exponentially fast.

To finish the computation of n_1 we note that the value of μ can be determined from the self-consistency equation: $\mu = \sum_{k=1}^{\infty} k^\gamma n_k$ which implies

$$1 = \sum_{k=1}^{\infty} \prod_{j=1}^{k} \left(1 + \frac{\mu}{j^\gamma}\right)^{-1} \tag{4.2.4}$$

Problem. Prove (A) by showing that any subsequential limit μ of $M_\gamma(t)/t$ satisfies (4.2.4).

Case 2. $\gamma > 2$. Let a and b be the two vertices in G_1. The probability the new edge always attaches to a is

$$\prod_{m=1}^{\infty} \frac{m^\gamma}{m + m^\gamma} = \prod_{m=1}^{\infty} \left(1 + \frac{1}{m^{\gamma-1}}\right)^{-1} > 0$$

It seems natural to

Conjecture. *There is an n_0 so that for $n \geq n_0$ all new vertices attach to one vertex.*

In any case, this does not seem to be a good model for sex in Sweden.

Case 3. $1 < \gamma \leq 2$. As in Case 1, the key problem is to understand the asymptotics of $M_\gamma(t)$, and we aren't quite able to prove what we need

(B) $M_\gamma(t)/t^\gamma \to 1$ as $t \to \infty$.

It is easy to see, as Krapivsky, Redner, and Leyvraz (2000) show in their equation (11),

$$M_\gamma(t) = \sum_{k} k^\gamma N_k(t) \leq t^{\gamma-1} \sum_{k} k N_k(t) \leq 2t^\gamma$$

With a little work this can be improved to

$$M_\gamma(t) < t + t^\gamma \tag{4.2.5}$$

Proof. We claim that the maximum of $\sum_k k^\gamma N_k(t)$ subject to $\sum_k N_k(t) = t+1$ and $\sum_k k N_k(t) = 2t$ is achieved by $N_1(t) = t$ and $N_t(t) = 1$. To prove this, suppose that the maximum is achieved by a configuration with $\sum_{j=2}^{t-1} N_j(t) > 0$. In this case we must have $N_t(t) = 0$ or we would have $\sum_k k N_k(t) > 2t$, and we must have $\sum_{j=2}^{t} N_j(t) \geq 2$ or we would have $\sum_k k N_k(t) < 2t$. We either have $1 < i < j < t$ with $N_i(t) > 0$ and $N_j(t) > 0$ or $1 < i = j < t$ with $N_i(t) \geq 2$. In either case when $i \leq j$

$$i^\gamma - (i-1)^\gamma = \int_{i-1}^{i} \gamma x^{\gamma+1} \, dx < \int_{j}^{j+1} \gamma x^{\gamma+1} \, dx = (j+1)^\gamma - j^\gamma$$

Rearranging we have $i^\gamma + j^\gamma < (i-1)^\gamma + (j+1)^\gamma$ or in words we can increase M_γ by spreading i and j further apart, a contradiction that proves (4.2.5). ∎

For a bound in the other direction we note that for $k \geq 2$

$$N_k(k) = \frac{(k-1)^\gamma}{M_\gamma(k-1)} N_{k-1}(k-1) = \prod_{j=2}^{k-1} \frac{j^\gamma}{M_\gamma(j)} N_2(2)$$

If $M_\gamma(t)/t^\gamma \to r < 1$ then we would get a contradiction to $N_k(k) \leq 1$. If you are a physicist the proof of (B) is complete. Unfortunately, as a mathematician I cannot rule out the paranoid delusion that $\limsup M_\gamma(t)/t^\gamma = 1$ and $\liminf M_\gamma(t)/t^\gamma < 1$, so I will regard (B) as an assumption waiting to be proved.

Once one believes (B), the rest is easy.

$$N_1(t) - N_1(t-1) = 1 - \frac{N_1(t-1)}{M_\gamma(t-1)}$$

Iterating, using $N_1(s) \leq s$ for $s \geq 2$, and the asymptotics for $M_\gamma(t)$:

$$N_1(t) - N_1(2) \geq t - 2 - \sum_{s=2}^{t-1} \frac{s}{M_\gamma(s)} \sim t$$

When $k = 2$ we have

$$N_2(t+1) - N_2(t) = \frac{1}{M_\gamma(t)} [N_1(t) - 2^\gamma N_2(t)] \leq \frac{t}{M_\gamma(t)}$$

The right-hand side $\sim t^{1-\gamma}$ so

$$\limsup_{t\to\infty} t^{-(2-\gamma)} N_2(t) \leq \limsup_{t\to\infty} t^{-(2-\gamma)} \sum_{s=1}^{t} s^{1-\gamma} = 1/(2-\gamma)$$

This calculation shows that discarding $2^\gamma N_2(t)$ from the previous equation had no affect on the limit, and we have

$$N_2(t)/t^{2-\gamma} \to 1/(2-\gamma)$$

Repeating the last argument for $k > 2$, we first get an upper bound on $N_k(t)$ from

$$N_k(t+1) - N_k(t) \leq \frac{(k-1)N_{k-1}(t)}{M_\gamma(t)}$$

Then we use the upper bound to conclude

$$N_k(t+1) - N_k(t) \sim \frac{(k-1)N_{k-1}(t)}{M_\gamma(t)}$$

and it follows by induction that

$$N_k(t) \sim c_k t^{1+(k-1)(1-\gamma)}$$

Noting that the exponent is < 0 when $k > \gamma/(\gamma - 1)$, it is natural to

Conjecture. *If $\gamma > 1$ then there is one vertex with degree of order $\sim t$. All of the others have degrees $O(1)$. There are only $O(1)$ vertices with degree $> \gamma/(\gamma - 1)$.*

Oliveira and Spencer (2005) have proved this conjecture and in addition have results on the geometry of the limiting graph.

Having found that powers change the behavior too much, we now consider

Case 4. Let $f(j) = a + j$ where $a > -1$. The small change of adding an a will make a significant change in the behavior. This time it is easy to find the asymptotics of

$$M_f(t) = \sum_j f(j)N_j(t) = N_1(t) + aN_0(t) = 2t + a(t+1)$$

so $M_f(t)/t \to \mu = 2 + a$. Let $n_k = \lim N_k(t)/t$. By calculations in Case 1, given in (4.2.2) and (4.2.3), $n_1 = \mu/(\mu + f(1))$ and

$$n_k = \prod_{j=2}^{k} \frac{f(j-1)}{\mu + f(j)} n_1 = \frac{\mu}{f(k)} \prod_{j=1}^{k} \frac{f(j)}{\mu + f(j)} = \frac{\mu}{a+k} \prod_{j=1}^{k} \left(1 + \frac{\mu}{a+j}\right)^{-1}$$

To determine the asymptotic behavior of n_k we note that

$$\log \prod_{j=1}^{k} \left(1 + \frac{\mu}{a+j}\right)^{-1} \sim - \sum_{j=1}^{k} \frac{\mu}{a+j} \sim -\mu \log k$$

so the product is $\sim k^{-\mu} = k^{-(2+a)}$. Recalling the $\mu/(a+k)$ out front the power is $k^{-(3+a)}$. Since $a > -1$ this means we can achieve any power > 2.

The models in this section are new, but they have close relatives that are very old.

Yule (1925) was interested in understanding the distribution of the number of species of a given genus. For this he introduced the following model:

- A genus starts with a single species. New species in the genus arise according to the Yule process in which individual gives birth at rate λ.
- Separately from within each genus a new genus appears at constant rate μ.

This is similar to the Case 4 model in that if we look at the embedded jump chain when there are k genera and a total of ℓ species then a genus with j species will get a new species with probability $\lambda j/(\lambda \ell + \mu k)$, while a new genus will be added with probability $\mu k/(\lambda \ell + \mu k)$.

It is well known that the number of individuals Z_t in a Yule process at time t has a geometric distribution

$$P(Z_t = n) = e^{-\lambda t}(1 - e^{-\lambda t})^{n-1} \quad \text{for } n \geq 1$$

From the second assumption the number of genera will grow exponentially at rate μ, so if we pick a random genus extant today, the time since its first appearance will have an exponential(μ) distribution. Thus the distribution of the number N of species in a random genus will be

$$p(n) = \int_0^\infty \mu e^{-\mu t} e^{-\lambda t}(1 - e^{-\lambda t})^{n-1} \, dt$$

letting $\alpha = \mu/\lambda$ and recognizing this as a beta integral we have

$$p(n) = \alpha \Gamma(1 + \alpha) \cdot \frac{\Gamma(n)}{\Gamma(n + 1 + \alpha)}$$

where Γ is the usual gamma function. From this it follows that $p(n) \sim Cn^{-1-\alpha}$ as $n \to \infty$. Since $\alpha > 0$ we can achieve any power > 1. For more on this see Aldous (2001).

Simon (1955) considered the following model of word usage in books, which he also applied to scientific publications, city sizes, and income distribution. Let $X_i(t)$ be the number of words that have appeared exactly i times in the first t words. He assumed that (i) there is a constant probability α that the $(t + 1)$th word is a word that has not appeared in the first t words, and (ii) if the $(t + 1)$th word is not new, the probability of a word is proportional to the number of times it has been used. Writing $N_i(t) = E X_i(t)$ we have

$$N_1(t + 1) - N_1(t) = \alpha - \frac{1 - \alpha}{t} N_1(t)$$
$$N_k(t + 1) - N_k(t) = \frac{1 - \alpha}{t}[(k - 1)N_{k-1}(t) - kN_k(t)] \quad \text{for } k > 1$$

Let $n_i = \lim N_i(t)/t$. Using Lemma 4.1.1 and noting $c = \alpha$, $b = 1 - \alpha$ gives

$$N_1(t)/t \to c/(1 + b) = \alpha/(2 - \alpha)$$

The second equation in this case is the same as the one in Case 1 with $f(i) = (1 - \alpha)i$ and $\mu = 1$, so using (4.2.2) and (4.2.3), we have $n_1 = \frac{\alpha}{2-\alpha}$ and for $k \geq 2$

$$n_k = \frac{\alpha}{(1-\alpha)k} \prod_{j=1}^{k} \frac{(1-\alpha)j}{1+(1-\alpha)j} = \frac{\alpha}{(1-\alpha)k} \prod_{j=1}^{k} \left(1 + \frac{1}{(1-\alpha)j}\right)^{-1}$$

Now $\log \prod_{j=1}^{k} \sim -\sum_{j=1}^{k} 1/(1-\alpha)j \sim -(\log k)/(1-\alpha)$, so the product is $\sim i^{-1/(1-\alpha)}$. Recalling the $1/(1-\alpha)i$ out front the power, is $i^{-(1+1/(1-\alpha))}$. Since $0 < \alpha < 1$ this means we can achieve any power > 2.

Back to the Recent Past. A number of researchers have modified the preferential attachment model and achieved powers $\neq 3$. Kumar et al. (2000) analyzed a "copying" model first introduced by Kleinberg et al. (1999). At each step one new vertex with out-degree d is added. To decide where the out-links should go they first choose a prototype vertex p uniformly at random from the current set of vertices. For $1 \leq i \leq d$ they use independent biased coin flips to determine whether the destination will be chosen uniformly at random with probability α or will be the end of the ith out link of p. Their Theorem 8 shows that the fraction of vertices of degree k converges to a limit p_k with $p_k \sim Ck^{-\beta}$ and $\beta = (2-\alpha)/(1-\alpha) \in (2, \infty)$.

Cooper and Frieze (2003a) considered a very general model in which at each step there is a probability α that an old node will generate edges and probability $1 - \alpha$ that a new node will be added. A new node generates i edges with probability p_i. With probability β terminal nodes are chosen uniformly at random, with probability $1 - \beta$ choices of terminal vertices are made according to degree. When an old node is chosen, with probability δ it is chosen uniformly at random, and with probability $1 - \delta$ with probability proportional to its degree. An old node generates i new edges with probability q_i. With probability γ terminal nodes are chosen uniformly at random, with probability $1 - \gamma$ choices of terminal vertices are made according to degree. In most cases a power law results. See pages 315–316 for the necessary notation and page 317 for the result itself. In section 6, Cooper and Frieze consider directed versions of the model. Bollobás, Borgs, Chayes, and Riordan (2003) considered a more general model where the in- and out-degree distributions have power laws with different exponents.

The references we have cited are just the tip of a rather large iceberg. More can be find in the cited sources and in the nice survey article by Mitzenmacher (2004).

4.3 Martingales and Urns

In the first two sections we have concentrated on the fraction of vertices with a fixed degree k. In this section we shift our attention to the other end of the spectrum and look at the growth of the degrees of a fixed vertex j. As in the previous section we use the fact that graph with n vertices and m edges are added at once, $G_n^{(m)}$ can be obtained by collapsing vertices in $G_{nm}^{(1)}$ to reduce to the case with $m = 1$.

Móri's Martingales. Móri (2002, 2005) has studied the Case 4 model from the previous section. As above he starts with G_1 consisting of two vertices, 0 and 1, connected by an edge, although as the reader will see the proof generalizes easily to other initial configurations. In his notation at the nth step, a new vertex is added and connected to an existing vertex. A vertex of degree k is chosen with probability $(k + \beta)/S_n$ where $\beta > -1$ and $S_n = 2n + (n + 1)\beta = (2 + \beta)n + \beta$ is the sum of the weights for the random tree with n edges and $n + 1$ vertices.

Let $X[n, j]$ be the weight (= degree $+\beta$) of vertex j after the nth step, let $\Delta[n + 1, j] = X[n + 1, j] - X[n, j]$, and let $\mathcal{F}_n$ denote the σ-field generated by the first n steps. If $j \leq n$ then

$$P(\Delta[n + 1, j] = 1|\mathcal{F}_n) = X[n, j]/S_n$$

From this, we get

$$E(X[n + 1, j]|\mathcal{F}_n) = X[n, j]\left(1 + \frac{1}{S_n}\right)$$

so $c_n X[n, j]$ will be a martingale if and only if $c_{n+1}/c_n = S_n/(1 + S_n)$.

Anticipating the definition of a larger collection of martingales we let

$$c[n, k] = \frac{\Gamma(n + \beta/(2 + \beta))}{\Gamma(n + (k + \beta)/(2 + \beta))} \quad n \geq 1, \ k \geq 0$$

where $\Gamma(r) = \int_0^\infty t^{r-1}e^{-t}\, dt$. For fixed k

$$c[n, k] = n^{-k/(2+\beta)}(1 + o(1)) \quad \text{as } n \to \infty \qquad (4.3.1)$$

Using the recursion $\Gamma(r) = (r - 1)\Gamma(r - 1)$ we have

$$\frac{c[n + 1, k]}{c[n, k]} = \frac{n + \beta/(2 + \beta)}{n + (k + \beta)/(2 + \beta)} = \frac{S_n}{S_n + k} \qquad (4.3.2)$$

and it follows that $c[n, 1]X[n, j]$ is a martingale for $n \geq j$. Being a positive martingale it will converge a.s. to a random variable ζ_j.

To study the joint distribution of the $X[n, j]$ we need another martingale.

Lemma 4.3.1. *Let $r, k_1, k_2, \ldots k_r$ be positive integers, and $0 \leq j_1 < \cdots < j_r$ be nonnegative integers. Then*

$$Z[n, \vec{j}, \vec{k}] = c[n, \sum_i k_i] \prod_{i=1}^{r} \binom{X[n, j_i] + k_i - 1}{k_i}$$

is a martingale for $n \geq \max\{j_r, 1\}$.

Proof. By considering two cases (no change or $\Delta[n + 1, j] = 1$) it is easy to check that

$$\binom{X[n + 1, j] + k - 1}{k} = \binom{X[n, j] + k - 1}{k}\left(1 + \frac{k\Delta[n + 1, j]}{X[n, j]}\right)$$

Since at most one $X[n, j]$ can change this implies

$$\prod_{i=1}^{r}\binom{X[n + 1, j_i] + k_i - 1}{k_i} = \prod_{i=1}^{r}\binom{X[n, j_i] + k_i - 1}{k_i}\left(1 + \sum_{i=1}^{r}\frac{k_i\Delta[n + 1, j_i]}{X[n, j_i]}\right)$$

Since $P(\Delta[n + 1, j] = 1|\mathcal{F}_n) = X[n, j]/S_n$, using the definition of $Z[n, \vec{j}, \vec{k}]$ and taking expected value

$$E(Z[n + 1, \vec{j}, \vec{k}]|\mathcal{F}_n) = Z[n, \vec{j}, \vec{k}] \cdot \frac{c[n + 1, \sum_i k_i]}{c[n, \sum_i k_i]}\left(1 + \frac{\sum_i k_i}{S_n}\right) = Z[n, \vec{j}, \vec{k}]$$

the last following from (4.3.2). ∎

Being a nonnegative martingale $Z[n, \vec{j}, \vec{k}]$ converges. From the form of the martingale, the convergence result for the factors, and the asymptotics for the normalizing constants in (4.3.1), the limit must be $\prod_{i=1}^{r} \zeta_{j_i}^{k_i}/k_i!$. Our next step is to check that the martingale converges in L^1. To do this we begin by observing that (4.3.1) implies $c[n, m]^2/c[n, 2m] \to 1$ and we have

$$\binom{x + k - 1}{k}^2 = \left(\frac{(x + k - 1)\cdots x}{k!}\right)^2 \le \frac{(x + 2k - 1)\cdots x}{2k!} \cdot \binom{2k}{k}$$

From this it follows that

$$Z[n, \vec{j}, \vec{k}]^2 \le C_k Z[n, \vec{j}, 2\vec{k}]$$

so our martingales are bounded in L^2 and hence converge in L^1. Taking $r = 1$ we have for $j \ge 1$

$$E\zeta_j^k/k! = \lim_{n\to\infty} EZ[n, j, k] = EZ[j, j, k] = c[j, k]\binom{k + \beta}{k}$$

while $E\zeta_0^k = E\zeta_1^k$.

Let M_n denote the maximal degree in our random tree after n steps, and for $n \ge j$ let $M[n, j] = \max\{Z[n, i, 1] : 0 \le i \le j\}$. Note that $M[n, n] = c[n, 1](M_n + \beta)$. Define $\mu(j) = \max\{\zeta_i : 0 \le i \le j\}$ and $\mu = \sup_{j\ge0} \zeta_j$.

Theorem 4.3.2. *With probability one,* $n^{-1/(2+\beta)}M_n \to \mu$.

Proof. Being a maximum of martingales, $M[n, n]$ is a submartingale. Using a trivial inequality and the fact that $Z[n, j, 1]^k$ is a submartingale for $k \geq 1$

$$EM[n, n]^k \leq \sum_{j=0}^{n} EZ[n, j, 1]^k \leq \sum_{j=0}^{\infty} E\zeta_j^k = k! \binom{k+\beta}{k} \sum_{j=0}^{\infty} c[j, k] < \infty$$

by (4.3.1) if $k > (2+\beta)$. Thus $M[n, n]$ is bounded in L^k for every integer $k > (2+\beta)$, and hence bounded and convergent in L^p for any $p \geq 1$.

Let $k > 2 + \beta$ be fixed. Then clearly

$$E(M[n, n] - M[n, j])^k \leq \sum_{i=j+1}^{n} EZ[n, i, 1]^k$$

The limit of the left-hand side is

$$E\left(\lim_{n\to\infty} n^{-1/(2+\beta)} M_n - \mu(j)\right)^k$$

while the right-hand side increases to

$$\sum_{i=j+1}^{\infty} E\zeta_i^k = k! \binom{k+\beta}{k} \sum_{i=j+1}^{\infty} c[i, k]$$

which is small if j is large. Hence $\lim_{n\to\infty} n^{-1/(2+\beta)} M_n = \mu$ as claimed. ∎

Urn Schemes. Berger, Borgs, Chayes, and Saberi (2005) have a different approach to preferential attachment models using Pólya urns. We begin by considering the $\beta = 0$ version of the model and start with G_1 consisting of one vertex, 1, and a self-loop, which we think of as an urn with two balls numbered 1. At each time $k \geq 2$, we draw one ball from the urn, then return it with a copy of the ball chosen and a new ball numbered k. In graph terms we add a new vertex numbered k, pick another vertex j with probability proportional to its degree and draw an edge from k to j.

At time k there are $2k$ balls in the urn. If we ignore balls with numbers $m > k$ and only increase time when a ball with number $j \leq k$ is drawn then we have Polya's urn scheme. If we consider balls numbered k to be red and those with $j < k$ to be black then we start with $r = 1$ red balls and $b = 2k - 1$ black balls. The probability that the first n_1 draws are red balls and the next $n_2 = n - n_1$ are black balls is

$$\frac{r(r+1)\cdots(r+n_1-1)\cdot b(b+1)\cdots(b+n_2-1)}{(r+b)(r+b+1)\cdots(r+b+n-1)}$$

A little thought reveals that if S is any subset of $\{1, 2, \ldots n\}$ with $|S| = n_1$ then in the probability that red balls are drawn at time S and black balls at $\{1, 2, \ldots n\} - S$ the same terms appear in the numerator but in a different order. Since the answer depends on the number of balls of each type drawn, but not on the order, then the sequence of draws is exchangeable.

By de Finetti's theorem, we know that the sequence of draws must be mixture of i.i.d. sequences. By the formula above, the probability of n_1 red and $n_2 = n - n_1$ black in the first n draws is

$$p_n(n_1) = \frac{n!}{n_1!n_2!} \cdot \frac{(r+n_1-1)!}{(r-1)!} \cdot \frac{(b+n_2-1)!}{(b-1)!} \cdot \frac{(r+b-1)!}{(r+b+n-1)!}$$

$$= \frac{(b+r-1)!}{(r-1)!(b-1)!} \cdot \frac{(r+n_1-1)!}{n_1!} \cdot \frac{(b+n_2-1)!}{n_2!} \cdot \frac{n!}{(r+b+n-1)!}$$

The product of the last three terms is

$$\frac{(n_1+r-1)(n_1+r-2)\cdots(n_1+1)(n_2+b-1)(n_2+b-2)\cdots(n_2+1)}{(n+r+b-1)(n+r+b-2)\cdots(n+1)}$$

If $n_1/n \to x$ then the last expression $\sim x^{r-1}(1-x)^{b-1}/n$ since r and b are fixed and there are $(r-1)+(b-1)$ terms in the numerator versus $r+b-1$ in the denominator. From this it follows that

$$p_n(n_1) \sim \frac{(b+r-1)!}{(r-1)!(b-1)!} x^{r-1}(1-x)^{b-1} \cdot \frac{1}{n}$$

so the mixing measure is beta(r, b).

One can also derive the last result by considering a Yule process in which each individual gives birth at rate 1. If we consider the times at which the population increases by 1 then the probability the new individual is of type j is proportional to the number of individuals of type j so we recover the urn scheme. If we start the Yule process with one individual at time 0 then the number at time t has a geometric distribution with mean e^t, that is,

$$P(Y(t) = n) = (1 - e^{-t})^{n-1} e^{-t} \quad \text{for } n \geq 1$$

From this it follows that $e^{-t}Y(t) \to \xi$ where ξ is exponential with mean one. If we consider the $2k$ initial balls to be founders of different families, then the limiting fraction of balls of type k (i.e., members of family $2k$) is

$$\frac{\xi_{2k}}{\xi_1 + \cdots + \xi_{2k}} = \text{beta}(1, 2k - 1)$$

where the ξ_i are independent exponentials. For more details see Section 9.1 of Chapter V of Athreya and Ney (1972).

Consider now the process with $\beta > 0$. Since the initial condition is a self-loop, the sum of the weights is $2n + \beta n$ and we can reformulate the graph model as: when the new vertex k is added the other endpoint is a vertex chosen uniformly at random with probability $\alpha = \beta/(\beta + 2)$ and with probability $1 - \alpha = 2/(2 + \beta)$ vertices are chosen with probability proportional to their degrees.

Again we will ignore balls with numbers $m > k$ and only increase time when a vertex with number $j \leq k$ is chosen for attachment. At time k the sum of the weights of the first k vertices is $2k + \beta k$. Let $T_0 = k$ and let T_m be the mth time after time k at which a vertex with degree $j \le k$ is chosen for attachment. If $d_j(m)$ is the degree of j after the attachment at time T_m then the probability k is chosen rather than some $j < k$ is

$$\frac{d_k(m-1) + \beta}{2k + (m-1) + k\beta}$$

Letting $r = 1$ and $b = 2k - 1$ as above, the probability that k is always chosen in the first n_1 draws and never in the next $n_2 = n - n_1$ is

$$\frac{(r + \beta) \cdots (r + n_1 - 1 + \beta) \cdot (b + \beta(k-1)) \cdots (b + n_2 - 1 + \beta(k-1))}{(r + b + \beta k) \cdots (r + b + n - 1 + \beta k)}$$

A little thought reveals that if S is any subset of $\{1, 2, \ldots n\}$ with $|S| = n_1$ then in the probability that k is chosen at T_m with $m \in S$ and not when $m \in \{1, 2, \ldots n\} - S$ the same terms appear in the numerator but in a different order. Since the answer depends on the number of balls of each type drawn, but not on the order, then the sequence of draws is again exchangeable.

By de Finetti's theorem, we know that it must be mixture of i.i.d. sequences. The probability of n_1 red and $n_2 = n - n_1$ black in the first n draws is

$$\frac{n!}{n_1! n_2!} \cdot \frac{\Gamma(r + n_1 + \beta)}{\Gamma(r + \beta)} \cdot \frac{\Gamma(b + n_2 + (k-1)\beta)}{\Gamma(b + (k-1)\beta)} \cdot \frac{\Gamma(r + b + k\beta)}{\Gamma(r + b + n + k\beta)}$$

$$= \frac{\Gamma(r + b + k\beta)}{\Gamma(r + \beta)\Gamma(b + (k-1)\beta)} \cdot \frac{\Gamma(r + n_1 + \beta)}{n_1!} \cdot \frac{\Gamma(b + n_2 + (k-1)\beta)}{n_2!}$$

$$\times \frac{n!}{\Gamma(r + b + n + k\beta)}$$

Now $\Gamma(n + c)/n! \sim n^{c-1}$ so if $n_1/n \to x$ then the expression above

$$\sim \frac{\Gamma(r + b + k\beta)}{\Gamma(r + \beta)\Gamma(b + (k-1)\beta)} x^{r+\beta-1}(1 - x)^{b+(k-1)\beta-1} \cdot \frac{1}{n}$$

and the mixing measure is beta$(r + \beta, b + (k - 1)\beta)$, where $r = 1$ and $b = 2k - 1$.

None of this long-winded discussion appears in Berger, Borgs, Chayes, and Saberi (2005), who do not give a proof of their Lemma 3.1, which gives the construction we are about to describe. Unfortunately they have the parameters of the beta distribution wrong by a little bit. They define u by $\alpha = u/(u + 1)$ in contrast to our $\alpha = \beta/(\beta + 2)$, and in the case $m = 1$ introduce ψ_k for $k \ge 2$ that are independent beta$(1 + u, 2k + ku)$. Correcting their typo we will take these variables to be beta$(1 + \beta, (2k - 1) + (k - 1)\beta)$. Let $\psi_1 = 1$ and for $1 \le k \le n$ let

$$\phi_k = \psi_k \prod_{j=k+1}^{n} (1 - \psi_j)$$

It is immediate from the definition that $\sum_{k=1}^{n} \phi_k = 1$. For help with the definitions that are about to come look at the following picture of $n = 6$.

Let $\ell_k = \sum_{j=1}^{k} \phi_j$. For $a \in (0, 1)$ let $\kappa(a) = \min\{k : \ell_k \geq a\}$. In words, $\kappa(a)$ is the number of the interval in which a falls. In the picture $\kappa(a) = 4$.

To construct the random graph $G_n^{(1)}$, we start with $G_1^{(1)}$ a self-loop. For $k \geq 2$, let U_k be independent random variables uniform on $[0, 1]$. When k is added we draw an edge to $j = \kappa(U_k \ell_{k-1})$. To see that this has the right distribution, note that (i) for fixed k if we condition on $(\phi_1/\ell_k, \phi_2/\ell_k, \ldots, \phi_k/\ell_k)$ and only look at the times at which the degree of some vertex $j \leq k$ is chosen then the choices are i.i.d., and (ii) the distribution of $(\phi_1/\ell_k, \phi_2/\ell_k, \ldots, \phi_k/\ell_k)$ gives the mixing measure we need to produce the urn scheme that governs the growth of the degrees.

Our main interest here has been to bring out the connection between preferential attachment and Pólya urns. To paraphrase Berger, Borgs, Chayes, and Saberi (2005), the Pólya urn representation is a generalization of Bollobás and Riordan's random pairing representation (a.k.a. linearized chord diagrams mentioned at the beginning of Section 4.6, but not discussed in this monograph), which extends it to the model in which vertices can choose their neighbors uniformly at random with some probability. It can be used to give a proof of the result on the diameter of the preferential attachment graph to be discussed in Section 4.6. In addition, it provides a proof of the "expanding neighborhood calculation" which is an ingredient of their analysis of the contact process in Section 4.8.

We would like to thank Jason Schweinsberg for his help with the second half of this section.

4.4 Scale-Free Trees

The Barabási–Albert model with $m = 1$ was introduced in a different context and under a different name by Szymánski (1987). A tree on $\{1, 2, \ldots n\}$ is *recursive* if each vertex other than 1 is joined to exactly one earlier vertex. The simplest such tree is the *uniform random recursive tree* grown one vertex at a time by joining the new vertex to an old vertex chosen at random. A *plane-oriented recursive tree* is one with a cyclic order on the edges meeting each vertex. Suppose that a new vertex v is added to an existing vertex w with degree d. Then there are d different ways in which the new vertex can be added, depending on where it is inserted in the cycle. For example, in the picture below, there are three ways to connect a vertex to 2.

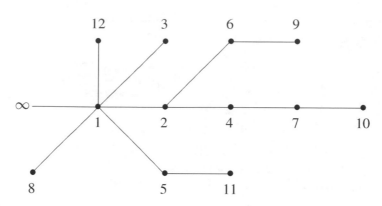

To explain the ∞, we start the process starts with a tree T_1 that has one vertex with an edge coming out of it connected to ∞, so that the degree of 1 is one initially. For $n \geq 2$, given T_{n-1}, vertex n is attached to vertex j with probability $d_{n-1}(j)/(2n-3)$, where $d_{n-1}(j)$ is the degree of j in T_{n-1}, so this is a version of the Barabási–Albert model with $m = 1$. Szabó, Alava, and Kertész (2002) were the first to study the shortest paths and load scaling (the use of edges by shortest paths) in this model, using mean-field calculations and simulation. Bollobás and Riordan (2004a) proved rigorous results, which in some cases corrected the earlier nonrigorous results. They based their calculations on a lemma that generalizes Lemma 4.6.4, and gives the probability a graph S occurs as a subgraph of T_n, see Corollary 22 in Bollobás and Riordan (2003). Their calculations are somewhat lengthy, so in this section, we will take a new approach to understanding the shape of the tree.

We begin with a study of the distances of the vertices from the root. Consider a branching random walk on the positive integers in which particles do not die, each particle gives birth at rate 1, and a particle at k gives birth to one offspring at k and to one at $k + 1$. To relate this to the Barabási–Albert model, the number of particles at k, $Z_t(k)$ is the sum of the degrees of the vertices at distance k from the root. When a change occurs, we pick a vertex x with probability proportional to its degree and connect a new vertex y. If x is at distance k from the root then y is at distance $k + 1$, and we have added one to the total degrees at k and $k + 1$.

Let Z_t be the total number of particles at time t. $(d/dt)EZ_t = 2EZ_t$ so $EZ_t = e^{2t}$. To see how these particles are distributed in space, let $m(k, t) = EZ_t(k)$. It is easy to see that $m(0, 0) = 1$ with $m(k, 0) = 0$ otherwise, and

$$\frac{d}{dt}m(k, t) = m(k - 1, t) + m(k, t)$$

To solve this let $p_t(x, y)$ be the transition probability of the random walk that jumps from x to $x + 1$ at rate 1. To check that $m(k, t) = e^{2t} p_t(0, k)$ is the solution to our equation, we differentiate and use the Kolmogorov equation for the random walk to check

$$\frac{d}{dt}e^{2t}p_t(0,k) = 2e^{2t}p_t(0,k) + e^{2t}(p_t(0,k-1) - p_t(0,k))$$
$$= e^{2t}(p_t(0,k-1) + p_t(0,k)) \qquad (4.4.1)$$

Since $EZ_t = e^{2t}$ and $e^{-2t}Z_t \to W$ almost surely, we are interested in the behavior of the system at time $t = (1/2)\log n$ when $Z_t = O(n)$. Let S_t be a Poisson random variable with mean t. It is clear from the description of the random walk that $p_t(0,x) = P(S_t = x)$. Taking $x = 0$, $p_t(0,0) = e^{-t}$ and $EZ_t(0) = e^t$. The particles at 0 are a branching process so $e^{-t}Z_t(0)$ converges almost surely to a limit $W(0)$. Taking $t = (1/2)\log n$ we see that the degree of the root is $O(n^{1/2})$ in agreement with the results in the previous section.

Since the Poisson has mean t and standard deviation $\sqrt{t}$, we expect that when the graph has n vertices most vertices will be at distance approximately $(1/2)\log n$ from the root. Using the local central limit theorem for the Poisson it follows that if $k = (1/2)\log n + x\sqrt{(1/2)\log n}$ then

$$EZ_{(1/2)\log n}(k) \sim \frac{n}{\sqrt{\pi \log n}}e^{-x^2/2} \qquad (4.4.2)$$

However, more is true. Results of Asmussen and Kaplan (1976) imply that

$$Z_{(1/2)\log n}(k)/EZ_{(1/2)\log n}(k) \to W \qquad \text{a.s.} \qquad (4.4.3)$$

where W is the limit of $e^{-2t}Z_t$.

To study the height of the tree, we use large deviations.

$$E\exp(\theta S_t) = \sum_{k=0}^{\infty} e^{-t}\frac{t^k}{k!}e^{\theta k} = \exp(t(e^\theta - 1))$$

Markov's inequality implies that if $a > 1$ and $\theta > 0$ then

$$P(S_t > at) \le e^{-\theta at}\exp(t(e^\theta - 1)) = \exp(t(-\theta a + e^\theta - 1))$$

To optimize the right-hand side we differentiate the exponent with respect to θ and set $-a + e^\theta = 0$ to conclude that the best choice of $\theta = \log a$. In this case we get

$$P(S_t > at) \le \exp(t[-a\log a + a - 1])$$

so in the branching random walk the expected number of particles in (a, ∞)

$$EZ_t(at, \infty) = e^{2t}P(S_t > at) \le \exp(t[1 - a\log a + a]) \qquad (4.4.4)$$

Changing variables $a = 1/\gamma$, the right-hand side is 1 when

$$1 - \frac{1}{\gamma}\log\left(\frac{1}{\gamma}\right) + \frac{1}{\gamma} = 0$$

Solving gives $\gamma + 1 = \log(1/\gamma)$ which means $\gamma e^{\gamma+1} = 1$. Let $a_0 = 1/\gamma_0$ where $\gamma_0 e^{\gamma_0+1} = 1$. If $a > a_0$, $EZ_t(at, \infty) \to 0$ exponentially fast in t, so eventually the right-most particle is $\leq a_0 t$. Biggins (1976) has shown that this upper bound is sharp, so the height of the tree $\sim a_0 t$ almost surely. This conclusion agrees with the result of Pittel (1994).

The large deviations result can be used to determine the size of $EZ_t(at)$ for $0 < a < a_0$. Following Biggins (1979), we define $\phi(\theta) = \exp(e^\theta - 1)$ and note that

$$\sum_k e^{\theta k} Z_t(k)/\phi(\theta)^t$$

is a nonnegative martingale and hence converges to a limit $W(\theta)$. To see which θ to choose, recall that $\theta_a = \log a$ optimizes the upper bound. Results on pages 20 and 21 of Biggins paper imply that for $0 < a < a_0$

$$Z_t(at)/EZ_t(at) \to W(\theta_a) \qquad \text{a.s.} \tag{4.4.5}$$

and show us that the number of vertices at different distances from the root, are up to a constant factor given by the expected value. To compute the expected value, and to sharpen we obtained earlier in (4.4.4), we use Stirling's formula $(at)! \sim (at)^{at} e^{-at} \sqrt{2\pi at}$ to conclude

$$EZ_t(at) = e^{2t} e^{-t} \frac{t^{at}}{(at)!} \sim \exp(t[1 - a \log a + a])/\sqrt{2\pi at} \tag{4.4.6}$$

Loads. Suppose that a vertex v has $c(v)$ descendants. Then the component of $T_n - \{v\}$ containing the root has $n - 1 - c(v)$ vertices so the number of shortest paths that use v is $c(v)[n - 1 - c(v)] + O(c(v)^2)$ so if $c(v) = o(n)$ then the load will be $\approx nc(v)$. To study the distribution of $c(v)$ we let $N_n(c)$ be the number of vertices in T_n with c descendants and note that $\lambda_{n,c} = E(N_n(c))$ satisfies for $c \geq 1$

$$\lambda_{n,c} = \left(1 - \frac{2c+1}{2n-3}\right) \lambda_{n-1,c} + \frac{2c-1}{2n-3} \lambda_{n-1,c-1} \tag{4.4.7}$$

To check this, note that if the number of descendants is c then v and its descendants are a tree S with $c + 1$ vertices and hence c edges. v is joined to its parent by a single edge, so the sum of the degrees of vertices in S is $2c + 1$, out of the total degree of $2n - 3$ in T_{n-1}. For $c = 0$ the only difference is that the new vertex always has 0 descendants so

$$\lambda_{n,0} = \left(1 - \frac{1}{2n-3}\right) \lambda_{n-1,0} + 1 \tag{4.4.8}$$

The last two equations have the form of those studied in Lemmas 4.1.1 with $c = 1$, $b = 1/2$, and 4.1.2 with $g = p_{c-1}(2c-1)/2$ and $b = (2c+1)/2$, so we conclude that $\lambda_{n,c}/n \to p_c$ where

$$p_0 = \frac{c}{1+b} = \frac{1}{1+1/2} = 2/3$$

$$p_c = \frac{g}{1+b} = p_{c-1}\frac{(2c-1)/2}{1+(2c+1)/2} = p_{c-1}\frac{2c-1}{2c+3}$$

Solving the recursion we have

$$p_c = \frac{2}{(2c+1)(2c+3)}$$

which agrees with the more precise result given in (2) of Bollobás and Riordan (2004a):

$$\lambda_{n,c} = \frac{2n-1}{(2c+1)(2c+3)} \quad \text{for } c \leq n-2$$

and $\lambda_{n,n-1} = 1$. To check this, note that $\lambda_{n-1,n-1} = 0$ so taking $c = n-1$ in (4.4.7) gives

$$\lambda_{n,n-1} = \frac{2(n-1)-1}{2n-3}\lambda_{n-1,n-2} = \lambda_{n-1,n-2}$$

When $1 \leq c \leq n-2$, (4.4.7) says

$$\lambda_{n,c} = \left(1 - \frac{2c+1}{2n-3}\right)\frac{2n-3}{(2c+1)(2c+3)} + \frac{2c-1}{2n-3}\cdot\frac{2n-3}{(2c-1)(2c+1)}$$

$$= \frac{2n-3}{(2c+1)(2c+3)} - \frac{1}{2c+3} + \frac{1}{2c+1} = \frac{2n-1}{(2c+1)(2c+3)}$$

Finally, when $c = 0$

$$\lambda_{n,0} = \frac{2n-4}{2n-3}\cdot\frac{2n-3}{3} + 1 = \frac{2n-1}{3}$$

Pairwise Distances. If $c \geq \epsilon n$ then

$$\lambda_{n,c} \leq \frac{2n}{(2\epsilon n)^2} = \frac{1}{2\epsilon^2 n}$$

so if we let $B = \{v : c(v) \geq \epsilon n\}$ then

$$E|B| = \sum_{c=\epsilon n}^{n} \lambda_{n,c} \leq \frac{1}{2\epsilon^2}$$

and it follows from Chebyshev's inequality that

$$P(|B| > \epsilon^{-3}) < \epsilon/2$$

If $v \in B$ then all of its ancestors are also in B, so B is connected subtree containing the root, and all of the components of $T_n - B$ have size $\leq \epsilon n$. If we pick two vertices x and y at random from $T_n - B$ then with probability $\geq 1 - 2\epsilon$ they are in different components of $T_n - B$. When this occurs the shortest path between these two vertices must pass through B, and hence the length of that path is almost the sum of the distances of x and y from the root. Using (4.4.3) and (4.4.2) we see that the individual distances are Poisson with mean $(1/2) \log n$. Since the vertices are chosen independently the sum is Poisson with mean $\log n$, which in turn is approximately normal with mean $\log n$ and variance $\log n$.

The last conclusion matches (4) in Bollobás and Riordan (2004a). Let E_k be the expected number of shortest paths of length k in T_n. They have also calculated that if $k / \log n = \alpha$ is bounded above and below by constants strictly between 0 and e

$$E_k = \Theta\left(n^{1-\alpha \log \alpha + \alpha} / \sqrt{\log n}\right)$$

There is a striking resemblance between the result we obtain by combining (4.4.5) and (4.4.6) and setting $t = (1/2) \log n$

$$Z_{(1/2)\log n}((a/2)\log n) \sim W(\theta_a) n^{(1/2)[1-a \log a+a]} / \sqrt{\pi a \log n}$$

Comparing the two formulas suggests that many of the paths of length k come from connecting two vertices that are at a distance $k/2$ from the root.

Remark. Goh, Kahng, and Kim (2001) and Goh et al. (2002) have studied load and "betweenness centrality" on power law random graphs with $2 < \gamma < 3$.

4.5 Distances: Power Laws $2 < \beta < 3$

Consider a Newman–Strogatz–Watts random graph with $p_k = k^{-\beta}/\zeta(\beta)$ where $2 < \beta < 3$ and $\zeta(\beta) = \sum_{k=1}^{\infty} k^{-\beta}$ is the constant need to make the sum 1. In this case, $q_{k-1} = kp_k/\mu = k^{1-\beta}/\zeta(\beta - 1)$, so the mean is infinite and the tail of the distribution

$$Q_K = \sum_{k=K}^{\infty} q_k \sim \frac{1}{\zeta(\beta - 1)(\beta - 2)} K^{2-\beta} \qquad (4.5.1)$$

To study the average distance between two randomly chosen points, we will first investigate the behavior of the branching process in order to figure out what to guess. The power $0 < \beta - 2 < 1$, and q_k is concentrated on the nonnegative integers so q_k is in the domain of attraction of a one-sided stable law with index $\alpha = \beta - 2$. To explain this let $X_1, X_2, \ldots$ be i.i.d. with distribution q_k, and let $S_n = X_1 + \cdots + X_n$.

To understand how S_n behaves, for $0 < a < b < \infty$, let

$$N_n(a, b) = |\{m \le n : X_m/n^{1/\alpha} \in (a, b)\}|$$

Let $B_\alpha = 1/\{\zeta(\beta - 1)(2 - \beta)\}$. For each m the probability $X_m \in (an^{1/\alpha}, bn^{1/\alpha})$ is

$$\sim \frac{1}{n} B_\alpha (a^{-\alpha} - b^{-\alpha})$$

Since the X_m are independent, $N_n(a, b) \Rightarrow N(a, b)$ has a Poisson distribution with mean

$$B_\alpha(a^{-\alpha} - b^{-\alpha}) = \int_a^b \frac{\alpha B_\alpha}{x^{\alpha+1}}\, dx \qquad (4.5.2)$$

If we interpret $N(a, b)$ as the number of points in (a, b) the limit is a Poisson process on $(0, \infty)$ with intensity $\alpha B_\alpha x^{-(\alpha+1)}$. There are finitely many points in (a, ∞) for $a > 0$ but infinitely many in $(0, \infty)$.

The last paragraph describes the limiting behavior of the random set

$$\mathcal{X}_n = \{X_m/n^{1/\alpha} : 1 \le m \le n\}$$

To describe the limit of $S_n/n^{1/\alpha}$, we will "sum up the points." Let $\epsilon > 0$ and

$$I_n(\epsilon) = \{m \le n : X_m > \epsilon n^{1/\alpha}\}$$
$$\hat{S}_n(\epsilon) = \sum_{m \in I_n(\epsilon)} X_m \qquad \bar{S}_n(\epsilon) = S_n - \hat{S}_n(\epsilon)$$

$I_n(\epsilon) =$ the indices of the "big terms," that is, those $> \epsilon n^{1/\alpha}$ in magnitude. $\hat{S}_n(\epsilon)$ is the sum of the big terms, and $\bar{S}_n(\epsilon)$ is the rest of the sum.

The first thing we will do is show that the contribution of $\bar{S}_n(\epsilon)$ is small if ϵ is small. To do this we note that

$$E\left(\frac{X_m}{n^{1/\alpha}}; X_m \le \epsilon n^{1/\alpha}\right) = B_\alpha \sum_{k=1}^{\epsilon n^{1/\alpha}} \frac{k^{2-\beta}}{n^{1/\alpha}} \sim B_\alpha \frac{\epsilon^{3-\beta}(n^{1/\alpha})^{2-\beta}}{3-\beta}$$

Since $\beta - 2 = \alpha$ multiplying on each side by n gives

$$E(\bar{S}_n(\epsilon)/n^{1/\alpha}) \to B_\alpha \epsilon^{3-\beta}/(3-\beta) \qquad (4.5.3)$$

If $Z = \text{Poisson}(\lambda)$ then

$$E(\exp(itaZ)) = \sum_{k=0}^{\infty} e^{-\lambda} \frac{e^{itak}\lambda^k}{k!} = \exp(\lambda(e^{ita} - 1))$$

Dividing (ϵ, ∞) into small strips, using independence of the number of points in different strips, and passing to the limit gives

$$E \exp(it\hat{S}_n(\epsilon)/n^{1/\alpha}) \to \exp\left(\int_\epsilon^\infty (e^{itx} - 1)\frac{\alpha B_\alpha}{x^{\alpha+1}}\, dx\right) \qquad (4.5.4)$$

Now $e^{itx} - 1 \sim itx$ as $t \to 0$ and $\alpha < 1$ so combining (4.5.3) and (4.5.4) and letting $\epsilon \to 0$ slowly (see (7.6) in Chapter 2 of Durrett (2004) for more details) we have

$$E \exp(it S_n / n^{1/\alpha}) \to \exp\left(\int_0^\infty (e^{itx} - 1) \frac{\alpha B_\alpha}{x^{\alpha+1}} \, dx \right)$$

This shows $S_n / n^{1/\alpha}$ has a limit. The limit is the one-sided stable law with index α, which we will denote by Γ_α

Branching Process. The next result and its proof come from Davies (1978).

Theorem 4.5.1. *Consider a branching process with offspring distribution ξ with $P(\xi > k) \sim B_\alpha k^{-\alpha}$ where $\alpha = \beta - 2 \in (0, 1)$. As $n \to \infty$, $\alpha^n \log(Z_n + 1) \to W$ with $P(W = 0) = \rho$ the extinction probability for the branching process.*

Proof. Now if $Z_n > 0$ then

$$Z_{n+1} = \sum_{i=1}^{Z_n} \xi_{n,i}$$

where the $\xi_{n,i}$ are independent and have the same distribution as ξ. We can write

$$\log(Z_{n+1} + 1) = \frac{1}{\alpha} \log(Z_n + 1) + \log Y_n$$

$$\text{where} \quad Y_n = \left(1 + \sum_{i=1}^{Z_n} \xi_{n,i} \right) \Big/ (Z_n + 1)^{1/\alpha}$$

Multiplying each side by α^n and iterating we have

$$\alpha^n \log(Z_{n+1} + 1) = \log(Z_1 + 1) + \alpha^n \log Y_n + \cdots + \alpha \log(Y_1)$$

As $n \to \infty$, Y_n converges to Γ_α. Straightforward but somewhat tedious estimates on the tail of the distribution of Y_n show that, see pages 474–477 of Davies (1978),

$$E\left(\sum_{n=1}^\infty \alpha^n \log^+ Y_n < \infty \right) \quad \text{and} \quad E\left(\sum_{n=1}^\infty \alpha^n \log^- Y_n < \infty \right)$$

This shows that $\lim_{n \to \infty} \alpha^n \log(Z_{n+1} + 1) = W$ exists.

It remains to show that the limit W is nontrivial. Davies has a complicated proof that involves getting upper and lower bounds on $1 - G_n(x)$ where G_n is the distribution of Z_n which allows him to conclude that if $J(x) = P(W \le x)$ then

$$\lim_{x \to \infty} \frac{-\log(1 - J(x))}{x} = 1$$

Problem. Find a simple proof that $P(W > 0) > 0$.

Once this is done it is reasonably straightforward to upgrade the conclusion to $J(0) = \rho$, where ρ is the extinction probability. To do this we use Theorem 2.1.9 to conclude that if we condition on nonextinction and look only at the individuals that have an infinite line of descent then the number of individuals in generation n, $\tilde{Z}_n$ is a branching process with offspring generating function

$$\tilde{\phi}(z) = \frac{\phi((1-\rho)z + \rho) - \rho}{1 - \rho}$$

where ρ is the extinction probability, that is, the smallest solution of $\phi(\rho) = \rho$ in $[0, 1]$.

Recalling that the convergence of $\phi(z) \to 1$ as $z \to 1$ gives information about the decay of the tail of the underlying distribution, it is easy to check that the new law is also in the domain of attraction of the stable law with index α. By the definition of the process $n \to \tilde{Z}_n$ is nondecreasing. Wait until the time $N = \min\{n : \tilde{Z}_n > M\}$. In order for $\alpha^n \log(\tilde{Z}_n + 1) \to 0$ this must occur for each of the M families at time N. However, we have already shown that the probability of a positive limit is $\delta > 0$, so the probability all M fail to produce a positive limit is $(1 - \delta)^M \to 0$ as $M \to \infty$. ∎

The double exponential growth of the branching process associated with the degree distribution $p_k = k^{-\beta}/\zeta(\beta)$ where $2 < \beta < 3$ suggests that the average distance between two members of the giant component will be $O(\log \log n)$. To determine the constant we note that our limit theorem says

$$\log(Z_t + 1) \approx \alpha^{-t} W$$

so $Z_t + 1 \approx \exp(\alpha^{-t} W)$. Replacing $Z_t + 1$ by n and solving gives $\log n = \alpha^{-t} W$. Discarding the W and writing $\alpha^{-t} = \exp(-t \log \alpha)$ we get

$$t \sim \frac{\log \log n}{\log(1/\alpha)} \tag{4.5.5}$$

While our heuristics have been for the branching processes associated with the Newman–Strogatz–Watts model, to prove a result about our power law graphs we will consider the Chung–Lu model. In the power law model the probability of having degree k is $p_k = k^{-\beta}/\zeta(\beta)$ where $\zeta(\beta) = \sum_{k=1}^{\infty} k^{-\beta}$. The probability of having degree $\geq K$ is $\sim BK^{-\beta+1}$ where $1/B = (\beta - 1)\zeta(\beta)$. Assuming the weights are decreasing we have

$$w_i = K \quad \text{when} \quad i/n = BK^{-\beta+1}$$

Solving gives

$$w_i = (i/nB)^{-1/(\beta-1)} \tag{4.5.6}$$

Theorem 4.5.2. *Consider Chung and Lu's power law graphs with* $2 < \beta < 3$. *Then the distance between two randomly chosen vertices in the giant component, H_n is asymptotically at most*

$$(2 + o(1)) \log \log n / (-\log(\beta - 2)).$$

Remarks

(i) Theorem 2.4.2 can be generalized to prove the existence of dangling ends, so the diameter is at least $O(\log n)$.

(ii) Chung and Lu's assumption that the average degree $d > 1$ is unnecessary. When $2 < \beta < 3$, $\bar{d} \to \infty$ so there is always a giant component. We use the first two steps of their proof as given in Chung and Lu (2002b, p. 15881; 2003, p. 98), with a small change in the first step to sharpen the constant, but replace the last two with a branching process comparison.

(iii) van der Hofstad, Hooghiemstra, and Znamenski (2007) have shown that Theorem 4.5.2 this gives the correct asymptotics for the Newman–Strogatz–Watts model and have shown fluctuations are $O(1)$. Note that the correct asymptotics are twice the heuristic that comes from growing one cluster to size n, but matches the guess that comes from growing two clusters to size $\sqrt{n}$.

(iv) Cohen and Havlin (2003) have an interesting approach to the lower bound. Start with the vertex of highest degree, which in this case is $d_1 = O(N^{1/(\beta-1)})$. Connect it to the vertices that have the next d_1 largest degrees and continue in the obvious way. The number of vertices within distance k of the origin obviously grows as quickly as any graph with the given degree sequence, but as Cohen and Havlin compute its diameter is at least $O(\log \log n)$.

Proof. Let $t = n^{1/\sqrt{\log \log n}}$. Note that t goes to ∞ slower than any power n and faster than any power of $\log n$. Consider the vertices $H_0 = \{i : w_i \geq t\}$. Looking at (4.5.6), the number of vertices in this class is asymptotically $n' = Bnt^{1-\beta}$. It is important to note that the expected degrees are assigned deterministically in the CL model, so the fraction of vertices here and $\mathrm{Vol}(G)$ below are not random. Recalling that i is connected to j with probability $w_i w_j / (\mathrm{Vol}(G))$ and for this range of β, $\mathrm{Vol}(G) \sim \mu n$ where μ is the average degree, we see that two vertices in H_0 are connected with probability $p \sim t^2 / \mu n$. $np \to \infty$ faster than $\log n$ so by Theorem 2.8.1 the probability H_0 is connected tends to 1. Using Theorem 2.8.6 we see that the diameter of H_0 is

$$\sim \frac{\log n'}{\log(n'p)} \sim \frac{\log n}{(3-\beta)\log t} = O(\sqrt{\log \log n})$$

To expand out from H_0 we will use the following easy result.

Lemma 4.5.3. *If $S \cap T = \emptyset$ and $\mathrm{Vol}(S)\mathrm{Vol}(T) \geq c\mathrm{Vol}(G)$ then the distance from S to T satisfies $P(d(S, T) > 1) \leq e^{-c}$.*

Proof. Since $1 - x \leq e^{-x}$

$$P(d(S, T) > 1) = \prod_{i \in S, j \in T} 1 - \frac{w_i w_j}{\mathrm{Vol}(G)} \leq \exp\left(-\sum_{i \in S, j \in T} \frac{w_i w_j}{\mathrm{Vol}(G)}\right) \leq e^{-c} \quad \blacksquare$$

The volume of $\{i : w_i > r\}$ is asymptotically

$$\sum_{i \leq Bnr^{1-\beta}} (i/nB)^{-1/(\beta-1)} \sim (nB)^{1/(\beta-1)} C(Bnr^{1-\beta})^{(\beta-2)/(\beta-1)} = C'nr^{2-\beta}$$

Using this result with S a single vertex gives the following.

Lemma 4.5.4. *Let $1 > \alpha > \beta - 2$, $0 < \epsilon < \alpha - (\beta - 2)$, and $r \geq (\log n)^{1/\epsilon}$. If $w_j \geq r^\alpha$ and $T = \{i : w_i > r\}$ then for large n*

$$P(d(\{j\}, T) > 1) \leq \exp(-Cr^{\alpha+(2-\beta)}/2\mu) \leq n^{-2}$$

Let $H_k = \{i : w_i > t^{\alpha^k}\}$ and suppose $t^{\alpha^k} \geq (\log n)^{1/\epsilon}$. Using the previous lemma we see that if $j \in H_{k+1}$ then with probability $\geq 1 - n^{-2}$, j is connected to a point in H_k. Since H_0 is connected it follows that each H_k with $k \leq \ell \equiv \inf\{i : t^{\alpha^i} < (\log n)^{1/\epsilon}\}$ is connected. Let $m = (\log \log n)/(-\log \alpha)$. This is chosen so that

$$\frac{\alpha^m \log n}{\sqrt{\log \log n}} < 1$$

and $t^{\alpha^m} < e$, so $\ell \leq m$.

H_0 has diameter $O(\sqrt{\log \log n})$, so at this point we have shown that H_ℓ is connected and has diameter smaller than $2m + O(\sqrt{\log \log n})$. To connect the remaining points we note that if n is large then $T = H_\ell$ has volume at least $(Cn/2)(\log n)^{(2-\beta)/\epsilon}$, so if S has volume $\geq (\log n)^{\alpha/\epsilon}$ then Lemma 4.5.3 implies

$$P(d(S, T) > 1) \leq \exp(-C(\log n)^{(\alpha-(\beta-2))/\epsilon}/2\mu) \leq n^{-2}$$

Thus the component of a site i will connect to H_ℓ if it reaches size $(\log n)^{\alpha/\epsilon}$. It is easy to see that collisions in the growth of the cluster to size $(\log n)^{\alpha/\epsilon}$ can be ignored, so for most points this is almost the same as survival of the associated branching process, and since the branching process grows doubly exponentially fast the amount of time required to reach this size is $O(\log \log \log n)$. $\quad \blacksquare$

Remark. When $1 < \beta < 2$ the transformed distribution has infinite mean. In this case van der Hofstad, Hooghiemstra, and Znamenski (2006) have shown that

$$\lim_{n \to \infty} P(H_n = 2) = 1 - \lim_{n \to \infty} P(H_n = 3) = p \in (0, 1)$$

4.6 Diameter: Barabási–Albert Model

Bollobás and Riordan (2004b) have performed a rigorous analysis of the diameter of the graphs produced by the Barabási–Albert model. They use a slight modification of the original model because it is equivalent to "linearized chord diagrams" that are produced by random pairings of $\{1, 2 \ldots 2n\}$. First, consider the case $m = 1$. They inductively define a sequence of directed random graphs G_1^t on $\{i : 1 \le i \le t\}$. Start with G_1^1 the graph with one vertex and one loop. Given G_1^{t-1} form G_1^t by adding the vertex t together with a directed edge from t to I where I is chosen randomly with

$$P(I = i) = \begin{cases} d_i^{t-1}/(2t - 1) & 1 \le i \le t - 1 \\ 1/(2t - 1) & i = t \end{cases} \qquad (4.6.1)$$

where d_i^{t-1} = the degree of i in G_1^{t-1}. In words, we consider the outgoing edge from t when we consider where to attach the other end of the edge. Note that each vertex will have out degree 1. To extend the definition to $m > 1$, use the definition above to define random graphs G_1^{mt} on $\{v_i : 1 \le i \le mt\}$ then combine the vertices $v_{(k-1)m+1}, \ldots v_{km}$ to make vertex k.

The main result to Bollobás and Riordan (2004b) is:

Theorem 4.6.1. *Let $m \ge 2$ and $\epsilon > 0$. Then with probability tending to 1, G_m^n is connected and*

$$(1 - \epsilon) \log n / (\log \log n) \le \text{diameter}(G_m^n) \le (1 + \epsilon) \log n / (\log \log n)$$

The case $m = 1$ is excluded because the upper bound is false in this case. As we saw in Section 4.4 the average pairwise distance is $O(\log n)$ in this case. Chung and Lu, see Theorem 4 on page 15880 in (2002b) or Theorem 2.6 on page 95 in (2003), proved that for their graphs with degree distribution $\sim Ck^{-3}$ the average distance is $O(\log n / (\log \log n))$.

Here we will content ourselves to prove the lower bound. To do this, we will consider G_1^N with $N = nm$. The idea behind the proof is to compare G_1^N with a random graph in which an edge from i to j is present with probability $c/\sqrt{ij}$. Let g_j be the vertex to which j sends an edge when it is added to the graph.

Lemma 4.6.2.

(a) If $1 \le i < j$ then $P(g_j = i) \le C_1(ij)^{-1/2}$.
(b) If $1 \le i < j < k$ then $P(g_j = i, g_k = i) \le C_2 i^{-1}(jk)^{-1/2}$.

Proof. Let $d_{t,i}$ be the degree of i in G_1^t. From the definition

$$P(g_t = i | G_1^{t-1}) = \begin{cases} d_{t-1,i}/(2t-1) & i < t \\ 1/(2t-1) & i = t \end{cases}$$

so if $t > i$ we have

$$E(d_{t,i} | G_1^{t-1}) = \left(1 + \frac{1}{2t-1}\right) d_{t-1,i}$$

or taking expected value

$$E d_{t,i} = \left(1 + \frac{1}{2t-1}\right) E d_{t-1,i} \tag{4.6.2}$$

Letting $\mu_{t,i} = E d_{t,i}$ and noting $\mu_{i,i} = 1 + 1/(2i-1)$ we have

$$\mu_{t,i} = \prod_{s=i}^{t} \left(1 + \frac{1}{2s-1}\right)$$

Taking logarithms and using $\log(1+x) \le x$ we have

$$\log \mu_{t,i} \le \sum_{s=i}^{t} \frac{1}{2s-1}$$

$\sum_{s=i+1}^{t} 1/(2s-1)$ is an approximating sum for $\int_{2i}^{2t} dx/x$ in which subdivisions have width 2 and the function is evaluated at the midpoint. Since $1/x$ is convex, it follows that

$$\sum_{s=i}^{t} \frac{1}{2s-1} \le \frac{1}{2i-1} + \frac{1}{2}(\log t - \log i) \le 1 + \frac{1}{2}\log(t/i)$$

Since $e \le 3$,

$$\mu_{t,i} \le 3\sqrt{t/i} \tag{4.6.3}$$

and it follows that

$$P(g_j = i) = \frac{\mu_{j-1,i}}{2j-1} \le 3\sqrt{\frac{j-1}{i}} \frac{1}{2j-1} \le \frac{3}{2}(ij)^{-1/2}$$

For the last inequality note that $\sqrt{j^2-j} \le (2j-1)/2$ (square both sides) implies

$$\frac{\sqrt{j-1}}{2j-1} \le \frac{1}{2\sqrt{j}} \tag{4.6.4}$$

For part (b) we need to consider second moments.

$$E(d_{t,i}^2|G_i^{t-1}) = d_{t-1,i}^2\left(1 - \frac{d_{t-1,i}}{2t-1}\right) + (d_{t-1,i}+1)^2\frac{d_{t-1,i}}{2t-1}$$

Expanding out the square in the second term:

$$= d_{t-1,i}^2\left(1 + \frac{2}{2t-1}\right) + \frac{d_{t-1,i}}{2t-1}$$

Taking expected value

$$E(d_{t,i}^2) = Ed_{t-1,i}^2\left(1 + \frac{2}{2t-1}\right) + \frac{Ed_{t-1,i}}{2t-1} \qquad (4.6.5)$$

Adding (4.6.2) and (4.6.5) and writing $\mu_{t,i}^{(2)}$ for $E(d_{t,i}^2) + E(d_{t,i})$ we have

$$\mu_{t,i}^{(2)} = \left(1 + \frac{2}{2t-1}\right)\mu_{t-1,i}^{(2)} = \left(\frac{2t+1}{2t-1}\right)\mu_{t-1,i}^{(2)}$$

Iterating we have

$$\mu_{t,i}^{(2)} = \prod_{s=i+1}^{t}\frac{2s+1}{2s-1}\mu_{i,i}^{(2)} = \frac{2t+1}{2i+1}\mu_{i,i}^{(2)} \qquad (4.6.6)$$

If $i < j$, $d_{j,i} = d_{j-1,i} + 1_{\{g_j=i\}}$ so

$$E(d_{j,i}1_{\{g_j=i\}}|G_1^{j-1}) = (d_{j-1,i}+1)\frac{d_{j-1,i}}{2j-1}$$

Taking expected value and using (4.6.6) gives

$$E(d_{j,i}1_{\{g_j=i\}}) = \frac{\mu_{j-1,i}^{(2)}}{2j-1} = \frac{\mu_{i,i}^{(2)}}{2i+1}$$

Conditioning on G_1^j and arguing as in the proof of (4.6.3)

$$E(d_{t,i}1_{\{g_j=i\}}) \leq 3\sqrt{\frac{t}{j}}\frac{\mu_{i,i}^{(2)}}{2i+1}$$

Now if $i < j < k$

$$P(g_j = i, g_k = i|G_1^{k-1}) = 1_{\{g_j=i\}}\frac{d_{k-1,i}}{2k-1}$$

Taking expected values gives

$$P(g_j = i, g_k = i) \leq \frac{3}{(2k-1)}\sqrt{\frac{k-1}{j}}\frac{\mu_{i,i}^{(2)}}{2i+1}$$

Using (4.6.4) to show $\sqrt{k-1}/(2k-1) \le 1/2\sqrt{k}$ and noting that $\mu_{i,i}^{(2)} = 2 + 3/(2i-1) \le 5$ the above is

$$\le \frac{15}{4} i^{-1}(jk)^{-1/2}$$

which completes the proof of (b). ∎

Lemma 4.6.3. *Let E and E' be events of the form*

$$E = \cap_{s=1}^{r}\{g_{j_s} = i_s\} \qquad E' = \cap_{s=1}^{r'}\{g_{j'_s} = i'_s\}$$

where $i_s < j_s$ and $i'_s < j'_s$ for all s. If the sets $\{i_1, \ldots, i_r\}$ and $\{i'_1, \ldots, i'_{r'}\}$ are disjoint then $P(E \cap E') \le P(E)P(E')$.

Proof. To prove this we use a different description of the process. We consider G_1^t as consisting of $2t$ half-edges numbered $1, 2, \ldots 2t$, each attached to a vertex from $1, \ldots t$. Given G_1^{t-1} we obtain G_1^t by adding $2t - 1$ to vertex t, picking h_t uniformly on $1, 2, \ldots 2t - 1$, attaching $2t$ to the vertex j that has the half-edge h_t, and drawing an edge from i to j. The next picture should help explain the algorithm.

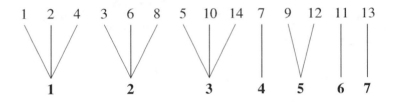

The choices that generated the graph were $h_2 = 2, h_3 = 3, h_4 = 6, h_5 = 5, h_6 = 9, h_7 = 10$ and the resulting graph is

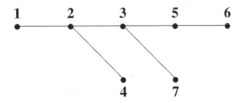

Note that the odd number $2i - 1$ always sits above vertex i. Because of this we have $g_j = i$ if $h_j = 2i - 1$. In the example, $h_3 = 3$, so $g_3 = 2$. When h_j is even it takes more work to figure out the value of g_j. If for some j' we have $h_j = 2j'$ and $h_{j'} = 2i - 1$ then $g_j = i$. In the example, $h_7 = 10$ and $h_5 = 5$ so $g_7 = 3$. In general, $g_j = i$ if

$$\exists s \ge 0, k_0 = j, k_1, \ldots k_s \text{ s.t. } h_{k_a} = 2k_{a+1} \text{ for } 0 \le a < s \text{ and } h_{k_s} = 2i - 1$$
$$(4.6.7)$$

Thus we can write E and E' as disjoint unions of events F_k and F'_ℓ that are intersections of events of the form (4.6.7). Now for each k and ℓ, the events are independent if no h_t occurs in both, or inconsistent since the common h_t's must have the same value and so cannot form a chain that leads to i and to $i' \neq i$. In either case we have $P(F_k \cap F'_\ell) \leq P(F_k)P(F'_\ell)$ and summing gives the desired result.   ∎

Lemma 4.6.4. *Let S be a graph on $\{1, 2, \ldots N\}$ in which each vertex is joined to at most two later vertices. Let $B = \max\{C_1, \sqrt{C_2}\}$. If $E(S)$ denotes the edges in S then*

$$P(S \subset G_1^N) \leq B^{|E(S)|} \prod_{ij \in E(S)} \frac{1}{\sqrt{ij}}$$

Proof. The event $S \subset G_1^N$ is the intersection of events $E_k = \{g_{j(k,1)} = \ldots g_{j(k,n_k)} = i_k\}$ where the i_k are distinct and each $n_k = 1, 2$. By Lemma 4.6.2

$$P(E_k) \leq B^{n_k} \prod_{\ell=1}^{n_k} (i_k j(k, \ell))^{-1/2}$$

By Lemma 4.6.3 the events E_k and $E_1 \cap \ldots \cap E_{k-1}$ are negatively correlated, so the desired result follows by induction.   ∎

Theorem 4.6.5. *Let $m \geq 1$ and let B be the constant in Lemma 4.6.4. Then with probability tending to 1,*

$$\text{diameter}\,(G_m^n) > \log n / \log(3Bm^2 \log n)$$

Proof. We will show that with probability tending to 1 the distance between n and $n - 1$ is greater than $L = \log n / \log(3Bm^2 \log n)$. Consider a self-avoiding path $V = v_0, v_1, \ldots v_\ell$. Recalling that G_m^n is obtained by identifying the vertices of G_1^{mn} in groups of m, this corresponds to a graph S that consists of edges $x_t y_{t+1}$, $t = 0, 1, \ldots \ell - 1$ with $\lceil x_t/m \rceil = \lceil y_t/m \rceil = v_t$, where $\lceil x \rceil$ rounds up to the next integer. Each vertex in S has degree at most two, so Lemma 4.6.4 implies S is present in G_1^{mn} with probability at most

$$B^\ell \prod_{t=0}^{\ell-1} \frac{1}{\sqrt{x_t y_{t+1}}} \leq B^\ell \prod_{t=0}^{\ell-1} \frac{1}{\sqrt{v_t v_{t+1}}} = \frac{B^\ell}{\sqrt{v_0 v_\ell}} \prod_{t=1}^{\ell-1} \frac{1}{v_t}$$

There are at most $m^{2\ell}$ graphs S that correspond to our path V so

$$P(V \subset G_m^n) \leq \frac{(Bm^2)^\ell}{\sqrt{v_0 v_\ell}} \prod_{t=1}^{\ell-1} \frac{1}{v_t}$$

The expected number of paths between $v_0 = n$ and $v_\ell = n - 1$ is at most

$$\leq \frac{(Bm^2)^\ell}{\sqrt{n(n-1)}} \sum_{1 \leq v_1, \ldots, v_{\ell-1} \leq n} \prod_{t=1}^{\ell-1} \frac{1}{v_t} = \frac{(Bm^2)^\ell}{\sqrt{n(n-1)}} \left(\sum_{v=1}^{n} \frac{1}{v} \right)^{\ell-1}$$

$$\leq \frac{(Bm^2)^\ell}{n-1} (1 + \log n)^{\ell-1} \leq \frac{(2Bm^2)^\ell}{n} (\log n)^{\ell-1} \qquad (4.6.8)$$

Now $\ell \leq L = \log n / \log(3Bm^2 \log n)$ implies $(3Bm^2 \log n)^\ell \leq n$ so the above is

$$\leq (2/3)^\ell (\log n)^{-1}$$

Summing over $\ell \leq L$ we find that the expected number of paths of length at most L joining the vertices n and $n-1$ tends to 0, which completes the proof. ∎

Problem. Can you derive Theorem 4.6.1 by considering the branching process with offspring distribution $q_{k-1} = k^{-2}/\zeta(2)$. There is a large literature on branching processes with infinite mean, but it does not seem to be useful for concrete examples like this one. It is my guess that $\log(1 + Z_t)/t \log t \to 1$.

4.7 Percolation, Resilience

One of the most cited properties of scale-free networks is that they are robust to random damage, but vulnerable to malicious attack. These questions were studied by Albert, Jeong, and Barabási (2000), Callaway, Newman, Strogatz, and Watts (2000), and Cohen, Erez, ben-Avraham, and Havlin (2000, 2001). Here, for the nonrigorous results we mostly follow Sections 11.3–11.5 of Dorogovtsev and Mendes (2002).

Random damage is modeled as a random removal of a fraction f of the nodes. If we recall our discussion of epidemics on networks with fixed degree distributions and identify the fraction $p = 1 - f$ edges that are retained with those that transmit the disease, (3.5.1) implies that the condition for the existence of giant component is

$$\nu p > 1 \qquad (4.7.1)$$

where $\nu = \sum_k k(k-1) p_k / \mu$ with $\mu = \sum_k k p_k$. In the scale-free case $p_k \sim C k^{-\gamma}$ with $2 < \gamma \leq 3$, $\nu = \infty$ and hence $p_c = 0$.

Our next objective is to compute the size of the giant component as a function of p for power laws with $2 < \gamma \leq 3$. According to Theorem 3.5.1, we want to find the smallest fixed point $\hat{G}_1(\xi) = \xi$ in $[0, 1]$ where $\hat{G}_1(z) = G_1(1 - p + zp)$ and G_1 is the generating function of the size-biased distribution $q_{k-1} = k p_k / \mu$. To understand the behavior of the generating function as $z \to 1$, we will use the following Tauberian theorem, which is Theorem 5 in Section XIII.5 of Feller (1971). To state this result we need a definition: $L(t)$ varies slowly at infinity if for any $0 < x < \infty$, $L(tx)/L(t) \to 1$. This obviously holds if L converges to a positive limit. $L(t) = (\log t)^a$ is a more interesting example.

Lemma 4.7.1. *Let $q_n \geq 0$ and suppose that $Q(s) = \sum_{n=0}^{\infty} q_n s^n$ converges for $0 \leq s < 1$. If L varies slowly at ∞ and $0 \leq \rho < \infty$ then each of the two following relations implies the other*

$$Q(s) \sim (1-s)^{-\rho} L(1/(1-s)) \qquad \text{as } s \uparrow 1$$

$$q_0 + \cdots + q_n \sim n^{\rho} L(n)/\Gamma(\rho+1)$$

Note that the result only gives us conclusions for the sum of the q's. To see that one cannot hope to get a result for the q's rather than their sums, define $\tilde{q}_{2n} = q_{2n-1} + q_{2n}$, $\tilde{q}_{2n-1} = 0$, and note that

$$\tilde{Q}(s) \leq Q(s) \leq \tilde{Q}(s)/s$$

so the two generating functions have the same behavior at 1.

The Tauberian theorem requires that we work with quantities that diverge as $s \uparrow 1$ so we consider

$$G_1'(z) = \sum_k k(k-1) p_k z^{k-2}/\mu$$

If $p_k \sim ck^{-\gamma}$ with $2 < \gamma < 3$ then

$$\sum_{k=1}^{n} k(k-1) p_k/\mu \sim cn^{3-\gamma}$$

so our Tauberian theorem implies

$$G_1'(z) \sim c(1-z)^{\gamma-3}$$

Integrating gives

$$G_1(1) - G_1(1-x) \sim \int_0^x cu^{\gamma-3}\, du = c'x^{\gamma-2}$$

To solve $G_1(1 - p + px) = x$ we write

$$1 - x = G_1(1) - G_1(1-p+xp) \sim c(p(1-x))^{\gamma-2}$$

Rearranging we have

$$(1-x) \sim cp^{(\gamma-2)/(3-\gamma)}$$

This gives the asymptotics for the survival probability $1 - \xi$ for the homogeneous branching process. Since G_0 has finite mean

$$1 - G_0(\xi) \sim G_0'(1)(1-\xi)$$

If $p_k \sim ck^{-3}$ then

$$\sum_{k=1}^{n} k(k-1)p_k/\mu \sim c\log n$$

so our Tauberian theorem implies

$$G_1'(z) \sim c\log(1 - z)$$

Integrating gives

$$G_1(1) - G_1(1-x) \sim -\int_0^x c\log u\,du \sim cx\log(1/x)$$

To solve $G_1(1 - p + px) = x$ we write

$$1 - x = G_1(1) - G_1(1 - p + xp) \sim cp(1-x)\log(1/p(1-x))$$

Rearranging we have $1/cp = \log(1/p) + \log(1/(1-x))$. Since $\log(1/p) \ll 1/p$ we have

$$1 - x = \exp(-(1 + o(1))/cp)$$

This give the asymptotics for the survival probability $1 - \xi$ for the homogeneous branching process. Since G_0 has finite mean

$$1 - G_0(\xi) \sim G_0'(1)(1 - \xi)$$

Combining our calculations and labeling the result to acknowledge it is a calculation for the branching process rather than the random graph itself, we have:

Claim 4.7.2. *Given a degree distribution with $p_k \sim ck^{-\gamma}$ with $2 < \gamma \leq 3$, the critical percolation probability is 0. If $2 < \gamma < 3$ the size of the giant component $\sim cp^{(\gamma-2)/(3-\gamma)}$ as $p \to 0$. If $\gamma = 3$ the size is $\exp(-(1 + o(1))/cp)$.*

Bollobás and Riordan (2004c) and Riordan (2004) have done a rigorous analysis of percolation on the Barabási–Albert preferential attachment graph, which has $\beta = 3$.

Theorem 4.7.3. *Let $m \geq 2$ be fixed. For $0 < p \leq 1$ there is a constant c_m and a function*

$$\lambda_m(p) = \exp\left(-\frac{c_m(1 + o(1))}{p}\right)$$

so that with probability $1 - o(1)$ the size of largest component is $(\lambda_m(p) + o(1))n$ and the second largest is $o(n)$.

The case $m = 1$ in which the world is a tree is not very interesting. For any $p < 1$ the largest component is $o(n)$.

Intentional Damage. It would be difficult to identify the collection of vertices of a given size that would do the most damage to the network, so in simulation studies, intentional damage is usually modeled as removal of the vertices with degrees $k > k_0$, where k_0 is chosen so that the desired fraction of vertices f is eliminated. The removal of the most connected sites leads to the disappearance of the links attached to them. This is equivalent to the removal of links chosen at random with probability

$$q = \sum_{k=k_0+1}^{\infty} kp_k \bigg/ \mu$$

The truncated distribution has all moments finite so by (4.7.1) it will have a giant component if

$$\frac{\sum_{k=1}^{k_0} k(k-1)p_k}{\sum_{k=1}^{k_0} kp_k}(1-q) > 1$$

Rearranging gives the simple condition

$$\sum_{k=1}^{k_0} k(k-1)p_k > \frac{\sum_{k=1}^{k_0} kp_k}{1-q} = \mu$$

which is equation (3) of Dorogovtsev and Mendes (2001). Callaway, Newman, Strogatz, and Watts (2000) have computed threshold values for the distribution $p_k = k^{-\gamma}/\zeta(\gamma)$ when $\gamma = 2.4, 2.7, 3.0$. The values of f_c and the corresponding k_0 are:

γ	f_c	k_0
2.4	0.023	9
2.7	0.010	10
3.0	0.002	14

This shows that the network is fragile – the removal of a few percent of the nodes with the largest degrees destroys the giant component. However, in interpreting these results the reader should keep in mind that $p_k = k^{-\gamma}/\zeta(\gamma)$ has no giant component for $\gamma > 3.479$, see Aiello, Chung, and Lu (2000, 2001).

Again, Bollobás and Riordan (2004c) have done a rigorous analysis for the preferential attachment model. For them it is convenient to define intentional damage as removal of the first nc nodes.

Theorem 4.7.4. *Let $m \geq 2$ and $0 < c < 1$ be constant. If $c \geq (m-1)/(m+1)$ then with probability $1 - o(1)$ the largest component is $o(n)$. If $c < (m-1)/$*

$(m + 1)$ *then there is a constant $\theta(c)$ so that with probability $1 - o(1)$ the largest component is $\sim \theta(c)n$, and the second largest is $o(n)$.*

It is difficult to compare this with the previous result since $p_k \sim 2m(m + 1)k^{-3}$ as $k \to \infty$. However, the reader should note that even when $m = 2$ one can remove 1/3 of the nodes.

4.8 SIS Epidemic

In this section we will study the susceptible–infected–susceptible epidemic on scale-free networks. In this model infected individuals become healthy at rate 1 (and are again susceptible to the disease) while susceptible individuals become infected at a rate λ times the number of infected neighbors. In the probability literature this model is usually called the contact process. Harris (1974) introduced this model. See Liggett (1999) for an account of most of the results known for this model.

Pastor-Satorras and Vespignani (2001a, 2001b, 2002) have made an extensive study of this model using mean-field methods, see also Moreno, Pastor-Satorras, and Vespignani (2002). These authors define the model as what probabilists call the threshold contact process: "each susceptible node is infected at rate λ if it is connected to one or more infected nodes" (see p. 3200 in 2001a, page 2 of 2001b, bottom of page 1 in 2002). However, it is clear from equation (4.8.1) below that infection occurs at a rate λ times the number of infected neighbors. Indeed if they considered the threshold contact process, then the maximum birth rate at a site is λ versus a death rate of 1, so $\lambda_c \geq 1$, in contrast to the main result which shows that when the tail of the degree distribution is large enough $\lambda_c = 0$.

Mean-Field Theory. Let $\rho_k(t)$ denote the fraction of vertices of degree k that are infected at time t, and $\theta(\lambda)$ be the probability that a given link points to an infected vertex. If we make the mean-field assumption that there are no correlations then

$$\frac{d\rho_k(t)}{dt} = -\rho_k(t) + \lambda k[1 - \rho_k(t)]\theta(\lambda)$$

so the equilibrium frequency ρ_k satisfies

$$0 = -\rho_k + \lambda k[1 - \rho_k]\theta(\lambda) \qquad (4.8.1)$$

Solving we have

$$\rho_k = \frac{k\lambda\theta}{1 + k\lambda\theta}$$

Suppose p_k is the degree distribution in the graph. The probability that a given link points to a vertex of degree k is $q_k = kp_k/\mu$ where $\mu = \sum_j jp_j$, so we have the following self-consistent equation for θ:

$$\theta = \sum_k q_k \rho_k = \sum_k q_k \frac{k\lambda\theta}{1 + k\lambda\theta} \qquad (4.8.2)$$

In the Barabási–Albert model $p_k \sim ck^{-3}$, or in the continuous approximation $p(x) = 2m^2/x^3$ for $x \geq m$. The size-biased distribution has $q(x) = m/x^2$ for $x \geq m$ and (4.8.2) becomes

$$\theta = \int_m^\infty \frac{m}{x} \frac{\lambda\theta}{1 + \lambda\theta x} \, dx = m \int_m^\infty \frac{\lambda\theta}{x} - \frac{(\lambda\theta)^2}{1 + \lambda\theta x} \, dx$$

The two parts of the integrand are not integrable separately, but if we replace the upper limit of ∞ by M the integral is

$$m\lambda\theta\{\log M - \log m\} - m\lambda\theta\{\log(1 + \lambda\theta M) - \log(1 + \lambda\theta m)\}$$

The first and third terms combine to $-m\lambda\theta \log(\lambda\theta + 1/M)$ so letting $M \to \infty$ the integral is

$$-m\lambda\theta \left(\log m + \log(\lambda\theta) - \log(1 + \lambda\theta m)\right) = m\lambda\theta \log\left(1 + \frac{1}{m\lambda\theta}\right)$$

and the equation we want to solve is

$$1 = m\lambda \log\left(1 + \frac{1}{m\lambda\theta}\right)$$

Dividing by $m\lambda$ and exponentiating

$$e^{1/m\lambda} = 1 + \frac{1}{m\lambda\theta}$$

Solving for θ now we have

$$\theta = \frac{1}{m\lambda(e^{1/m\lambda} - 1)} = \frac{e^{-1/m\lambda}}{m\lambda}(1 - e^{-1/m\lambda})^{-1} \qquad (4.8.3)$$

in agreement with (12) in Pastor-Satorras and Vespignani (2001b). The fraction of occupied sites

$$\rho = \sum_k p_k \frac{k\lambda\theta}{1 + k\lambda\theta} \sim \lambda\theta\mu \qquad (4.8.4)$$

as $\lambda, \theta \to 0$ by the dominated convergence theorem.

In the formula for θ, m and λ appear as the product $m\lambda$. A little thought reveals that this will always be the case if we work with continuous variables, so we will for simplicity restrict our attention to the case $m = 1$. Turning to powers between 2 and 3, let $p(x) = (1 + \gamma)x^{-2-\gamma}$ for $x \geq 1$ and assume $0 < \gamma < 1$. In this case the size-biased distribution is $q(x) = \gamma x^{-1-\gamma}$ and (4.8.2) becomes

$$1 = \int_1^\infty \frac{\gamma}{x^\gamma} \frac{\lambda}{1 + \lambda\theta x}\, dx$$

The right-hand side is a decreasing function of θ that is ∞ when $\theta = 0$ and $\to 0$ when $\theta \to \infty$ so we know there is a unique solution. Changing variables $x = u/\lambda\theta$, $dx = du/(\lambda\theta)$ we have

$$1 = \lambda^\gamma \theta^{\gamma-1} \int_{\lambda\theta}^\infty \gamma u^{-\gamma} \frac{1}{1 + u}\, du$$

Since $\gamma < 1$ the integral on the right has a limit c_γ as $\lambda, \theta \to 0$. Rearranging we have

$$\theta \sim (c_\gamma \lambda^\gamma)^{1/(1-\gamma)} \tag{4.8.5}$$

in agreement with (22) in Pastor-Satorras and Vespignani (2001b). Again, the fraction of occupied sites

$$\rho = \sum_k p_k \frac{k\lambda\theta}{1 + k\lambda\theta} \sim \lambda\theta\mu \tag{4.8.6}$$

as $\lambda, \theta \to 0$ by the dominated convergence theorem.

Turning to powers larger than 3, let $p(x) = (1 + \gamma)x^{-2-\gamma}$ for $x \geq 1$ and assume $\gamma > 1$. Again, the size-biased distribution is $q(x) = \gamma x^{-1-\gamma}$ and (4.8.2) is

$$1 = \int_1^\infty \frac{\gamma}{x^\gamma} \frac{\lambda}{1 + \lambda\theta x}\, dx \tag{4.8.7}$$

However, now the integral converges when $\theta = 0$, so for a solution to exist we must have

$$\lambda > \lambda_c = 1 \Big/ \int_1^\infty \frac{\gamma}{x^\gamma}\, dx = \frac{\gamma - 1}{\gamma}$$

Letting $F(\lambda, \theta)$ denote the right-hand side of (4.8.7), we want to solve $F(\lambda, \theta) = 1$. If $\lambda > \lambda_c$, $F(\lambda, 0) = \lambda/\lambda_c > 1$. To find the point where $F(\lambda, \theta)$ crosses 1 we note that

$$\frac{\partial F}{\partial \theta} = -\int_1^\infty \frac{\gamma}{x^\gamma} \frac{\lambda^2 x}{(1 + \lambda\theta x)^2}\, dx$$

When $\gamma < 2$, $\partial F/\partial\theta \to \infty$ as $\theta \to 0$. Changing variables $y = \theta x$ the above becomes

$$-\int_\theta^\infty \frac{\gamma\theta^\gamma}{y^\gamma} \frac{\lambda^2 y/\theta}{(1 + \lambda y)^2} \frac{dy}{\theta} \sim -\theta^{\gamma-2} \int_0^\infty \frac{\gamma}{y^{\gamma-1}} \frac{\lambda^2}{(1 + \lambda y)^2}\, dy$$

Writing $c_{\gamma,\lambda}$ for the integral (which is finite) and integrating

$$F(\lambda, \theta) - F(\lambda, 0) \sim -c_{\gamma,\lambda}\theta^{\gamma-1}/(\gamma - 1)$$

Rearranging

$$\theta_c \sim \left((\gamma - 1) \frac{F(\lambda, 0) - 1}{c_{\gamma, \lambda}} \right)^{1/(\gamma - 1)}$$

Recalling $F(\lambda, 0) = \lambda/\lambda_c$, it follows that

$$\theta(\lambda) \sim C(\lambda - \lambda_c)^{1/(\gamma - 1)}$$

Thus the critical exponent $\beta = 1/(\gamma - 1) > 1$ when $1 < \gamma < 2$. When $\gamma > 2$, $\partial F/\partial \theta$ has a finite limit as $\theta \to 0$ and $\beta = 1$.

The mean-field calculations above will not accurately predict equilibrium densities or critical values (when they are positive). However, they suggest the following conjectures about the contact process on power law graph with degree distribution $p_k \sim Ck^{-\alpha}$.

- If $\alpha \leq 3$ then $\lambda_c = 0$
- If $3 < \alpha < 4$, $\lambda_c > 0$ but the critical exponent $\beta > 1$
- If $\alpha > 4$ then $\lambda_c > 0$ and the equilibrium density $\sim C(\lambda - \lambda_c)$ as $\lambda \downarrow \lambda_c$.

Rigorous Results. Berger, Borgs, Chayes, and Saberi (2005) have considered the contact process on the Barabási–Albert preferential attachment graph. They have shown that $\lambda_c = 0$ and more. Here $\Theta(f(\lambda))$ indicates a quantity that is bounded above and below by a constant times $f(\lambda)$ as $\lambda \to 0$.

Theorem 4.8.1. *For every $\lambda > 0$ there is an N so that for a typical sample of the scale-free graph of size $n > N$, and a vertex v chosen at random:*

(a) the probability that the infection starting from v will survive is

$$\lambda^{\Theta(1)}$$

(b) with probability $1 - O(\lambda^2)$ the probability that the infection starting from v will survive is

$$\lambda^{\Theta(\log(1/\lambda)/\log\log(1/\lambda))}$$

The main ideas that underlie the proof of this theorem are:

(i) if all degrees in a graph are significantly smaller than λ^{-1} then the disease will die out very quickly
(ii) if the virus reaches a vertex with degree significantly larger than λ^{-2} then the disease is likely to survive for a long time
(iii) for a typical vertex v, the largest degree of a vertex in a ball of radius k around v is, with high probability $(k!)^{\Theta(1)}$.

The first two conclusions imply that the survival of the disease boils down to whether it can reach a vertex of degree $\lambda^{-\Theta(1)}$, while the last conclusion implies that the closest vertex of degree $\lambda^{-\Theta(1)}$ is $\Theta(\log(1/\lambda)/\log\log(1/\lambda))$.

The first conclusion is easy to see. If all the degrees are $\leq (1/2)\lambda^{-1}$ then the number of infected sites can be bounded above by a branching process in which births occur at rate $1/2$ and deaths occur at rate 1, so starting with a single infected site the expected number of infected sites at time t is $\leq e^{-t/2}$. To establish the second conclusion, it is enough to consider a star-shaped graph in which a central vertex of degree $C\lambda^{-2}$ is connected to that many leaves, that is, vertices with degree 1. If the central vertex was always occupied then the outlying vertices would be independently occupied with probability $\lambda/(\lambda+1) \approx \lambda$, so there would be an average of $C\lambda^{-1}$ occupied vertices, and if the central vertex becomes vacant then the time until it becomes occupied again will be $O(1)$.

The last calculation indicates why degree $C\lambda^{-2}$ is necessary for prolonged survival. To prove it is sufficient, we will use the following result which is a version of Berger, Borgs, Chayes, and Saberi (2005) Lemma 5.3. We have added the assumption $k\lambda^2 \to \infty$ which seems necessary for the conclusion.

Lemma 4.8.2. *Let G be a star graph with center 0 and leaves $1, 2, \ldots k$. Let A_t be the set of vertices infected in the contact process at time t when $A_0 = \{0\}$. If $k\lambda^2 \to \infty$ then $P(A_{\exp(k\lambda^2/10)} \neq \emptyset) \to 1$.*

Proof. Write the state of the system as (m, n) where m is the number of infected leaves and $n = 1$ if the center is infected and 0 otherwise. To reduce to a one-dimensional chain we will ignore the times at which the second coordinate is 0. When the state is $(m, 0)$ with $m > 0$, the probability that the next event will be the reinfection of the center is $\lambda/(\lambda+1)$, so the number of leaf infections N that will die while the center is 0 has a shifted geometric distribution with success probability $\lambda/(\lambda+1)$, that is,

$$P(N = j) = \left(\frac{1}{\lambda+1}\right)^j \cdot \frac{\lambda}{\lambda+1} \quad \text{for } j \geq 0$$

There is of course the possibility that starting with m infected leaves, all of them will become healthy before the center is reinfected but this will not occur if $N < m$.

The second step is to modify the chain so that the infection rate is 0 when this number is $L = \lambda k/4$ or greater. In this case the number of infected leaves $\geq Y_t$ where

$$
\begin{array}{ll}
 & \text{at rate} \\
Y_t \to Y_{t-1} - 1 & \lambda k/4 \\
Y_t \to Y_{t+1} + 1 & 3\lambda k/4 \\
Y_t \to Y_t - N & 1
\end{array}
$$

To bound the survival time of this chain we will estimate the probability that starting from $L - 1$ it will return to L before hitting 0. During this time Y_t is random walk that jumps at rate $\lambda k + 1$ and with the following distribution

$$
\begin{array}{ll}
 & \text{with probability} \\
-1 & (\lambda k/4)/(\lambda k + 1) \\
+1 & (3\lambda k/4)/(\lambda k + 1) \\
-N & 1/(\lambda k + 1)
\end{array}
$$

At this point we diverge from the Berger, Borgs, Chayes, and Saberi (2005) proof and finish up using a martingale. If we let X be a random variable with the distribution given above then

$$
E e^{\theta X} = e^\theta \cdot \frac{3}{4} \cdot \frac{\lambda k}{\lambda k + 1} + e^{-\theta} \cdot \frac{1}{4} \cdot \frac{\lambda k}{\lambda k + 1}
$$
$$
+ \frac{1}{\lambda k + 1} \sum_{j=0}^{\infty} e^{-\theta j} \left(\frac{1}{\lambda + 1} \right)^j \cdot \frac{\lambda}{\lambda + 1}
$$

If $e^{-\theta}/(\lambda + 1) < 1$, the sum of the geometric series is

$$
\frac{\lambda}{\lambda k + 1} \cdot \frac{1}{1 + \lambda - e^{-\theta}}
$$

If we pick $\theta < 0$ so that $e^{-\theta} = 1 + \lambda/2$ then

$$
\frac{\lambda k + 1}{\lambda k} E e^{\theta X} = \frac{1}{1 + \lambda/2} \cdot \frac{3}{4} + (1 + \lambda/2) \cdot \frac{1}{4} + \frac{2}{\lambda k}
$$

If $\lambda \to 0$ and $\lambda k \to \infty$ then the right-hand side converges to $1/2$ and hence is eventually positive.

To estimate the hitting probability we note that if $\phi(x) = \exp(\theta x)$ then $\phi(Y_t)$ is a martingale. Let q be the probability that Y_t hits $(-\infty, 0]$ before returning to L. Since $\theta < 0$, we have $\phi(x) \geq \phi(0)$ for $x \leq 0$ and the optional stopping theorem implies that

$$
q\phi(0) + (1 - q)\phi(L) \leq \phi(L - 1)
$$

Solving and using the fact that $\phi(x) \geq 0$ is decreasing, we have

$$
q \leq \frac{\phi(L - 1) - \phi(L)}{\phi(0) - \phi(L)} \leq \frac{\phi(L) - 0}{\phi(0) - \phi(L)} = \frac{1}{e^{-\theta L} - 1}
$$

Recalling $\theta \sim -\lambda/2$ and $L = \lambda k/4$ with $\lambda^2 k \to \infty$ we see that when $\lambda^2 k$ is large

$$
q \leq (1/2)e^{-\lambda^2 k/8}
$$

At this point we have estimated the probability that the chain started at $L - 1$ will go to L before hitting 0. When at L the time until the next jump to $L - 1$

is exponential with mean $1/(L+1)$. At this point we have shown that with high probability the chain will return from $L-1$ to L, $e^{\lambda^2 k/9}$ times before hitting 0. Using the law of large numbers we see that with high probability that many returns from L to $L-1$ will take at least $e^{\lambda^2 k/9}/2(L+1)$ units of time and the proof of the lemma is complete. ∎

Recently Chatterjee and Durrett (2009) have shown that if $\alpha > 3$ then $\lambda_c = 0$.

5

Small Worlds

5.1 Watts and Strogatz Model

As explained in more detail in Section 1.3, our next model was inspired by the popular concept of "six degrees of separation," which is based on the notion that every one in the world is connected to everyone else through a chain of at most six mutual acquaintances. Now an Erdös–Rényi random graph for $n = 6$ billion people in which each individual has an average of $\mu = 42.62$ friends would have average pairwise distance $(\log n)/(\log \mu) = 6$, but would have very few triangles, while in social networks if A and B are friends and A and C are friends, then it is fairly likely that B and C are also friends.

To construct a network with small diameter and a positive density of triangles, Watts and Strogatz (1998) started from a ring lattice with n vertices and k edges per vertex, and then rewired each edge with probability p, connecting one end to a vertex chosen at random. This construction interpolates between regularity ($p = 0$) and disorder ($p = 1$). The disordered graph is not quite an Erdös–Rényi graph, since the degree of a node is approximately the sum of a Binomial(k,1/2) and an independent Poisson($k/2$). Let $L(p)$ be the distance between two randomly chosen vertices. Define the clustering coefficient $C(p)$ to be the fraction of connections that exist between the $\binom{k}{2}$ neighbors of a site.

Suppose that $n \gg k \gg \log n \gg 1$. Extrapolating from the results for Erdös–Rényi graphs, we know that the middle condition implies that the graph will be connected with high probability when $p = 1$ and the diameter will be asymptotically $(\log n)/(\log k)$. Considering the first two steps in the cluster growth branching process tells us that the clustering coefficient $C(1) \sim k/n$. At the other extreme of perfect order, since we can move distance $k/2$ in one step and the maximum distance is $n/2$, $L(0) \sim n/k$.

Our next step is to show $C(0) \to 3/4$. Suppose $k = 2j$. The pairs of points $-j \le y < x \le j$ form a triangle with vertices $(j, j-1)$, $(-(j-1), -j)$, and $(j, -j)$. The points below the line $x - y = j$ are not neighbors, and this is asymptotically 1/4 of the triangle. The next figure shows the situation when $j = 5$.

132

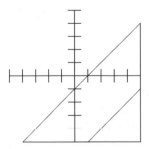

The next illustration, which is a copy of Figure 2 in Watts and Strogatz (1998), considers $n = 1000$ vertices and $k = 10$ neighbors, and shows that there is a broad interval of p over which $L(p)$ is almost as small as $L(1)$, yet $C(p)$ is far from 0. To see the reason for this, note that when a fraction $p = 0.01$ of the edges have been rewired, $C(p)$ has not changed by much, but the shortcuts have dramatically decreased the distance between sites.

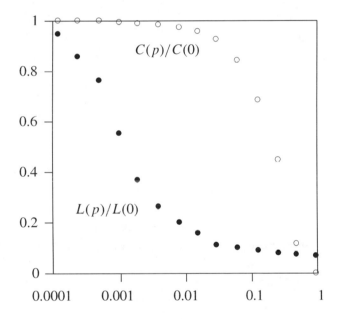

To look for the small world phenomenon in real graphs, Watts and Strogatz (1998) computed L and C for three examples: the collaboration graph of actors in feature films, the electrical power grid of the Western United States, and the neural network of the nematode worm *C. elegans*. Results are given in the next table and are compared to the values $L,$ and $C,$ for random graphs with the same number of vertices and average number of edges per vertex. As these results show the distances are similar to the random graphs in the first two cases, but 50% larger in the third. However, the clustering coefficients in the real graphs are significantly larger than in the random graphs, especially when the number of vertices is large in the case of film actors.

	L	L_r	C	C_r
C. elegans	2.65	2.25	0.28	0.05
Film actors	3.65	2.99	0.79	0.00027
Power grid	18.7	12.4	0.08	0.005

Bollobás and Chung Small World. Watts and Strogatz (1998) were not the first to notice that random long distance connections could drastically reduce the diameter. Bollobás and Chung (1988) added a random matching to a ring of n vertices with nearest neighbor connections and showed that the resulting graph had diameter $\sim \log_2 n$. This graph, which we will call the BC small world, is not a good model of a social network because (a) every individual has exactly three friends including one long-range neighbor, and (b) does not have any triangles, so it is locally tree like. These weaknesses, particularly the second, make it easier to study, so we will have a preference for this case throughout most of the chapter. In the section on epidemics and the final section on the contact process, we will include nonnearest neighbor connections. There, as in the other models considered in this chapter, the qualitative behavior is the same but the proofs are more difficult.

5.2 Path Lengths

In this section we are concerned with estimating the average path length between two randomly chosen sites in the small world, $\ell(n, p)$ as a function of the number of nodes n, the fraction of shortcuts p, and the range of interaction k. For this problem and the others we will consider below, we will consider Newman and Watts (1999) version of the model in which no edges are removed but one adds a Poisson number of shortcuts with mean $pkn/2$ and attaches then to randomly chosen sites.

To quote Albert and Barabási (2002), "it is now widely accepted that the characteristic path length obeys the general scaling form

$$\ell(n, p) \sim \frac{n}{k} f(pkn)$$

where $f(u)$ is a universal scaling function that obeys

$$f(u) = \begin{cases} 1/4 & \text{if } u \ll 1 \\ \ln(u)/u & \text{if } u \gg 1 \end{cases}"$$

Newman, Moore, and Watts (2000) have taken a "mean-field approach" to computing $\ell(n, p)$. They write differential equations for the number of sites within distance r of a fixed point and the number of clusters of occupied sites, assuming that gaps between clusters have the sizes of a randomly broken stick, that is, the result of putting that many i.i.d. uniforms in the unit interval. They conclude that

$$f(u) = \frac{1}{2\sqrt{u^2 + 2u}} \tanh^{-1}\left(\frac{u}{\sqrt{u^2 + 2u}}\right) \tag{5.2.1}$$

Simulations show that this formula agrees with simulations for small u or large u, but as expected, there is some disagreement when $u \approx 1$, see Figure 3 in Newman (2000). Using the identity

$$\tanh^{-1} y = \frac{1}{2} \log\left(\frac{1+y}{1-y}\right)$$

we have

$$\tanh^{-1}\left(\frac{u}{\sqrt{u^2 + 2u}}\right) = \frac{1}{2} \log\left(\frac{1 + u/\sqrt{u^2 + 2u}}{1 - u/\sqrt{u^2 + 2u}}\right)$$

Inside the logarithm the numerator $\rightarrow 2$ as $u \rightarrow \infty$. The denominator

$$= 1 - \frac{1}{\sqrt{1 + 2/u}} \approx 1 - \frac{1}{1 + 1/u} \approx 1/u$$

combining our calculations

$$f(u) \sim \frac{\log(2u)}{4u}$$

which matches (21) in Newman, Moore, and Watts (2000).

Barbour and Reinert (2001) have done a rigorous analysis of the average distance between points in a continuum model in which there is a circle of circumference L and a Poisson mean $L\rho/2$ number of random chords. The chords are the shortcuts and have length 0. The first step in their analysis is to consider an upper bound model that ignores intersections of growing arcs and that assumes each arc sees independent Poisson processes of shortcut endpoints. Let $S(t)$ be size, that is, the Lebesgue measure, of the set of points within distance t of a chosen point and let $M(t)$ be the number of intervals. Under our assumptions

$$S'(t) = 2M(t)$$

while $M(t)$ is a branching process in which there are no deaths and births occur at rate 2ρ.

$M(t)$ is a Yule process with births at rate 2ρ so $EM(t) = e^{2\rho t}$ and $M(t)$ has a geometric distribution

$$P(M(t) = k) = (1 - e^{-2\rho t})^{k-1} e^{-2\rho t} \tag{5.2.2}$$

Being a branching process $e^{-2\rho t} M(t) \rightarrow W$ almost surely. It follows from (5.2.2) that W has an exponential distribution with mean 1. Integrating gives

$$ES(t) = \int_0^t 2e^{2\rho s} \, ds = \frac{1}{\rho}(e^{2\rho t} - 1)$$

At time $t = (2\rho)^{-1}(1/2)\log(L\rho)$, $ES(t) = (L/\rho)^{1/2} - 1$. Ignoring the -1 we see that if we have two independent clusters run for this time then the expected number of connections between them is

$$\sqrt{\frac{L}{\rho}} \cdot \rho \cdot \frac{\sqrt{L/\rho}}{L} = 1$$

since the middle factor gives the expected number of shortcuts per unit distance and the last one is the probability a shortcut will hit the second cluster. The precise result is:

Theorem 5.2.1. *Suppose $L\rho \to \infty$. Let O be a fixed point of the circle, choose P at random, and let D be the distance from O to P. Then*

$$P\left[D > \frac{1}{\rho}\left(\frac{1}{2}\log(L\rho) + x\right)\right] \to \int_0^\infty \frac{e^{-y}}{1 + 2e^{2x}y}\, dy$$

Thus as in Theorem 3.4.1 the fluctuations are $O(1)$. To make a connection between the two results we note that the proof will show that the right-hand side is $E\exp(-2e^{2x}WW')$ where W and W' are independent exponentials.

Proof. To prove this we begin with a Poisson approximation result of Arratia, Goldstein, and Gordon (1990). Suppose $X_\alpha, \alpha \in I$ are Bernoulli random variables with $P(X_\alpha = 1) = p_\alpha$. Let $V = \sum_\alpha X_\alpha$, $\lambda = EV$, and Z be Poisson with mean λ. We are interested in conditions that imply V and Z are close in distribution. For each $\alpha \in I$ let $B_\alpha \subset I$ be a set that contains α. Intuitively, B_α is the neighborhood of dependence of X_α. Variables outside the neighborhood will be almost independent of X_α. Define

$$b_1 = \sum_{\alpha \in I} \sum_{\beta \in B_\alpha} p_\alpha p_\beta$$

$$b_2 = \sum_{\alpha \in I} \sum_{\beta \in B_\alpha, \beta \neq \alpha} E(X_\alpha X_\beta)$$

$$b_3 = \sum_{\alpha \in I} E|E(X_\alpha - p_\alpha | X_\gamma, \gamma \notin B_\alpha)|$$

Theorem 5.2.2. *Let $\mathcal{L}(V)$ be distribution of V. The total variation distance*

$$\|\mathcal{L}(V) - \mathcal{L}(Z)\| \leq 2\left((b_1 + b_2)\left(\frac{1 - e^{-\lambda}}{\lambda}\right) + b_3(1 \wedge 1.4\lambda^{-1/2})\right)$$

$$\leq 2(b_1 + b_2 + b_3)$$

To apply this suppose that we have intervals $I_1, \ldots, I_m$ with lengths $s_1, \ldots, s_m$ and intervals $J_1, \ldots, J_n$ with lengths $u_1, \ldots, u_n$ that are scattered independently and uniformly on a circle of circumference L. Let X_{ij} be the indicator of the event $I_i \cap J_j \neq \emptyset$ and $V = \sum_{i=1}^{m} \sum_{j=1}^{n} X_{ij}$.

$$p_{i,j} \equiv P(I_i \cap J_j \neq \emptyset) = (s_i + u_j)/L$$

so if we let $s = s_1 + \cdots + s_m$ and $u = u_1 + \cdots + u_n$ then

$$\lambda \equiv EV = \sum_{i=1}^{m} \sum_{j=1}^{n} (s_i + u_j)/L = (ns + mu)/L$$

We define $B_{i,j} = \{(i,k) : k \neq j\} \cup \{(\ell, j) : \ell \neq i\}$ so that if $(k, \ell) \notin B_{i,j}$ and $(k, \ell) \neq (i, j)$ then $X_{i,j}$ and $X_{k,\ell}$ are independent and hence $b_3 = 0$. If we let

$$Z_{i,j} = \sum_{(k,\ell) \in B_{i,j}} X_{k,\ell}$$

then we have

$$b_1 = \sum_{i,j} p_{i,j} E Z_{i,j} + \sum_{i,j} p_{i,j}^2$$

$$b_2 = \sum_{i,j} E(X_{i,j} Z_{i,j}) = \sum_{i,j} p_{i,j} E Z_{i,j}$$

since the $X_{i,j}$ are pairwise independent. To see this note that if $i \neq k$ and $j \neq \ell$ then $X_{i,j}$ and $X_{k,\ell}$ are clearly independent. To complete the proof now, it suffices to consider the case $i = k$ and $j \neq \ell$. However, in this situation even if we condition on the location of I_i the two events $I_i \cap J_j \neq \emptyset$ and $I_i \cap J_\ell \neq \emptyset$ are independent.

Let $r = \max_i s_i + \max_j u_j = L \max_{ij} p_{i,j}$. $E X_{i,j} \leq r/L$ so $E Z_{i,j} \leq (m+n-2) r/L$ and

$$\sum_{i,j} p_{i,j} E Z_{i,j} \leq \lambda (m + n - 2) r/L$$

The final term $\sum_{i,j} p_{i,j}^2 \leq (r/L) \sum_{i,j} p_{i,j} = r\lambda/L$ so $b_1 + b_2 \leq 2\lambda (m + n) r/L$. Using the first inequality in Theorem 5.2.2 with $1 - e^{-\lambda} \leq 1$, it follows that if Z is Poisson(λ)

$$\|\mathcal{L}(V) - \mathcal{L}(Z)\| \leq 4(m + n) r/L \tag{5.2.3}$$

Let $\tau_x = (2\rho)^{-1} \{(1/2) \log(L\rho) + x\}$. Consider two independent copies of the upper bound model starting from O and P and run until time τ_x. Let M_x and N_x be the number of intervals in the two processes, let s_x and u_x be the Lebesgue measure of the sets of points, and let $\hat{V}_x$ be the number of intersections. From (5.2.3) and $r \leq 4\tau_x$ it is immediate that

$$|P(\hat{V}_x = 0 | M_x, N_x, s_x, u_x) - \exp(-L^{-1}(N_x s_x + M_x u_x))| \leq 16(M_x + N_x)\tau_x/L \tag{5.2.4}$$

Taking expected value, then putting the expected value inside the absolute value

$$|P(\hat{V}_x = 0) - E\exp(-L^{-1}(N_x s_x + M_x u_x))| \leq \frac{16\tau_x}{L}E(M_x + N_x) \qquad (5.2.5)$$

Our next step is to estimate the number of collisions between the growing intervals in the upper bound process starting from O. Number the intervals I_j in the order in which they were created. Let $Y_{i,j} = 1\{I_i \cap I_j \neq \emptyset\}$ and $G_i = \{Y_{i,j} = 0$ for all $j < i\}$. Each interval I_i with $i > 1$ has a parent, $P(i)$, which was the source of the chord that started it. Let $H_1 = 0$ and

$$H_i = \begin{cases} 0 & \text{on } G_i \cap \{H_{P(i)} = 0\} \\ 1 & \text{otherwise} \end{cases}$$

$H_i = 1$ indicates an interval that is bad due to experiencing a collision or being a descendant of a bad interval.

Lemma 5.2.3. *If $P(Y_{i,j} = 1) \leq p$ for all i, j then $P(H_i = 1) \leq 2(i-1)p$.*

Proof. We prove the result by induction on i. The conclusion is clear for $i = 1$. $H_i = 1$ can occur for two reasons. The first is $P(G_i^c) \leq (i-1)p$. The second is that k is an ancestor of i and $H_k = 0$. Now since the intervals are numbered in order of their creation, their lengths are a decreasing function of their indices, and hence the probability j is the parent of i is a decreasing function on $1, \ldots i-1$. Iterating we see that if we follow the ancestry of i back until we first reach an interval $j \leq k$ then the probability we will end up at k is $\leq 1/k$. Using induction now

$$\sum_{k=1}^{i-1} P(H_k = 0, k \text{ is an ancestor of } i) \leq \sum_{k=1}^{i-1}(k-1)p/k \leq (i-1)p$$

which completes the proof of the lemma. ∎

Let V_x be the number of intersections in the real process in which intervals stop growing when they run into each other.

Lemma 5.2.4. *With the notations above we have*

$$P(\hat{V}_x \neq V_x) \leq \frac{32\tau_x^2}{L^2}E(M_x(M_x - 1)N_x)$$

Proof. Define $Y'_{i,j} = 1\{J_i \cap J_j \neq \emptyset\}$ and H'_i for the process starting at P as in Lemma 5.2.3 and recall that $X_{i,j} = 1\{I_i \cap J_j \neq \emptyset\}$. Since $V_x \leq \hat{V}_x$ are integer valued

$$P(\hat{V}_x \neq V_x) \leq E(\hat{V}_x - V_x) \leq E\left(\sum_{i=1}^{M_x}\sum_{j=1}^{N_x} X_{i,j}(1\{H_i = 1\} + 1\{H'_j = 1\})\right)$$

Conditioning on $M_x = m$, $N_x = n$ and on the lengths of the intervals, using the trivial observation that all intervals have length $\leq 2\tau_x$ and applying Lemma 5.2.3 we conclude

$$
\begin{aligned}
&E(X_{i,j} 1\{H_i = 1\} | M_x = m, N_x = n, s_1, \dots s_m, u_1, \dots u_n) \\
&\quad \leq (4\tau_x/L) P(H_i = 1 | M_x = m, N_x = n, s_1, \dots s_m, u_1, \dots u_n) \\
&\quad \leq (4\tau_x/L) \cdot 2(i-1)(4\tau_x/L)
\end{aligned}
$$

Noting $\sum_{i=1}^{k} 2(i-1) = k(k-1)$, combining this with a similar bound for $X_{i,j} 1\{H'_j = 1\}$ and using $E(N_x(N_x - 1)M_x) = E(M_x(M_x - 1)N_x)$ gives the desired result. ∎

Theorem 5.2.1 follows easily by combining (5.2.5) and Lemma 5.2.4. To do this we need to recall that if G has a geometric distribution with success probability q then $EG = 1/q$ and $E(G(G-1)) = (1-q)/q^2 \leq 1/q^2$. From this and the definition of $\tau_x = (2\rho)^{-1}\{(1/2)\log(L\rho) + x\}$ we have $EM_x \sim (L\rho)^{1/2} e^x$ and $EM_x(M_x - 1) \leq (L\rho)e^{2x}$. Using this with the cited results, writing $\ell_x = (1/2)\log(L\rho) + x$ and recalling $\tau_x = \ell_x/(2\rho)$ we have

$$
\begin{aligned}
&|P(V_x = 0) - E\exp(-L^{-1}(N_x s_x + M_x u_x))| \\
&\quad \leq \frac{16\tau_x}{L} 2(L\rho)^{1/2} e^x + \frac{32\tau_x^2}{L^2}(L\rho)^{3/2} e^{3x} \\
&\quad \leq \frac{16\ell_x}{(L\rho)^{1/2}} e^x + \frac{8\ell_x}{(L\rho)^{1/2}} e^{3x}
\end{aligned}
$$

which $\rightarrow 0$ if $x \leq (1/7)\log(L\rho)$.

To complete the proof we have to evaluate the expected value. Noting that

$$
\frac{S(t)}{M(t)} = \int_0^t \frac{M(r)}{M(t)} dr \rightarrow \int_0^\infty e^{-\rho s} ds = \rho^{-1}
$$

and recalling s_x and u_x are the total lengths of the intervals, we have

$$
L^{-1}(N_x s_x + M_x u_x) \sim 2(L\rho)^{-1} M_x N_x \rightarrow -2e^{2x} WW'
$$

where W and W' are independent exponential mean 1. The bounded convergence theorem implies

$$
E\exp(-L^{-1}(N_x s_x + M_x u_x)) \rightarrow E\exp(-2e^{2x} WW')
$$

If we condition on the value of W and use the formula for the Laplace transform of the exponential

$$
E\exp(-2e^{2x} WW') = E\left(\frac{1}{1 + 2e^{2x}W}\right) = \int_0^\infty \frac{1}{1 + 2e^{2x}y} e^{-y} dy
$$

which completes the proof of the theorem. ∎

5.3 Epidemics

In this section we will follow Moore and Newman (2000) and consider epidemic models on the small world, which are essentially percolation processes. There are two extremes: in the first all individuals are susceptible and there is a probability p that an infected individual will transmit the infection to a neighbor, in the second only a fraction p of individuals are susceptible but the disease is so contagious that if an individual gets infected all of their susceptible neighbors will become infected. In percolation terms, the first model is bond percolation while the second is site percolation. The qualitative properties of the model are similar. We will concentrate on the site percolation version since in that case it is possible to do the computations more explicitly.

To give our first nonrigorous derivation of the answer, we will introduce an infinite graph associated with the small world, that we call the "Big World." We begin with a copy of the integers, $\mathbb{Z}$. To each integer we attach a Poisson mean ρ long-range bonds that lead to a new copy of $\mathbb{Z}$ on which we repeat the previous construction. The first copy of $\mathbb{Z}$ we call level zero. The levels of other copies are equal to the number of long-range bonds we need to traverse to get to them. This structure appeared in an implicit way in the previous section: if we look at how the set of sites within distance n of 0 in the Big World grows then there are no collisions and each interval encounters an independent set of long-range connections.

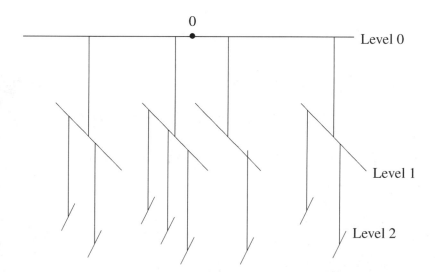

Site Percolation. To analyze the growth of the process, let $p_0(n)$ be the probability 0 is connected to n sites on Level 0. $p_0(0) = 1 - p$. The number of sites to the right of zero that can be reached has a geometric distribution with success probability $(1 - p)^k$, since it takes k consecutive closed sites to stop the percolation, and every time we can reach a new open site we can forget about the states of sites behind it.

(This is false for bond percolation when $k > 1$ and makes the calculations in that case much more difficult.)

Letting $q = (1 - p)^k$, the probability of reaching $j \geq 0$ sites on the right is $(1 - q)^j q$. Adding the sites reached on the left and noting that 0 has to be open to get the process started we have

$$p_0(n) - \sum_{j=0}^{n-1} p(1-q)^j q \cdot (1-q)^{n-1-j} q = np(1-q)^{n-1} q^2$$

Noting that the geometrics start at 0, so their means are $1/q - 1$, the mean number of sites reached on level 0 is

$$\nu \equiv \sum_n np_0(n) = p \cdot \frac{2 - q}{q}$$

Conditional on $N = n$ sites being reached on level 0 the number of longrange bonds M to level one copies will be Poisson with mean $n\rho$. Thus $E(M|N) = \rho N$ and $EM = \rho\nu$. Each level one copy reached starts an independent version of the original process. Thus if we let Z_k be the number of level k copies reached then Z_k is a branching process. The critical value for percolation occurs when $\rho\nu = 1$.

Bond Percolation, $k = 1$. This time 0 does not have to be open so

$$p_0(n) = \sum_{j=0}^{n-1} p^j(1-p) \cdot p^{n-1-j}(1-p) = np^{n-1}(1-p)^2$$

and the mean number of sites reached on level 0 is

$$\nu = 1 + 2p/(1-p) = (1+p)/(1-p)$$

This time the edges have to be open in order to reach the next level so $E(M|N) = p\rho N$ and $EM = p\rho\nu$. The critical value for percolation occurs when $p_c\rho\nu = 1$ or

$$\rho p_c \frac{1 + p_c}{1 - p_c} = 1$$

Solving we have $\rho p_c^2 + (\rho + 1)p_c - 1 = 0$ or

$$p_c = \frac{-(\rho + 1) + \sqrt{(\rho + 1)^2 + 4\rho}}{2\rho}$$

In order to check this result and to prepare for developments in section 5.5, we will now give another (nonrigorous) derivation of the bond percolation critical value based on the fact that, seen from a fixed vertex, the NW small world is locally tree like. Color vertices blue if they are reached by a long-range edge and red if they are reached by a short-range edge. Ignoring collisions the growth of the cluster is a two-type branching process with mean matrix

$$
\begin{array}{cc}
 & \begin{array}{cc} B & R \end{array} \\
\begin{array}{c} B \\ R \end{array} & \begin{array}{cc} \rho & 2 \\ \rho & 1 \end{array}
\end{array}
$$

The growth rate of this system is dictated by the largest eigenvalue of this matrix, which solves

$$0 = (\rho - \lambda)(1 - \lambda) - 2\rho = \lambda^2 - (\rho + 1)\lambda - \rho$$

Comparing with the previous quadratic equation we see that $p_c = 1/\lambda$. This is exactly what we should have expected since particles in generation n of the branching process are infected in the epidemic with probability p^n.

Rigorous Proof of Critical Values. Rather than take our usual approach of showing that the branching process accurately models the growth of the cluster, we will prove the result by reducing to a model with a fixed degree distribution. The reduction is based on the following picture.

If we only use the connections around the ring then we get connected components that have a geometric distribution with success probability r where $r = q$ for site percolation and $r = (1 - p)$ for bond percolation. In the case of site percolation we are ignoring closed sites, which are components of size 0. The distribution is a single geometric rather than the sum of two since we scan from left to right to find the successive components.

   Now each site in the cluster is connected to a Poisson mean $\lambda = \rho p$ number of edges. In bond percolation this comes from the fact that the edge has to be open to count. In the case of site percolation this comes from the fact that of the Poisson mean ρ edges, a fraction $(1 - p)$ are connected to sites that are closed. Collapsing the components to single vertices, they have degree $S_N = X_1 + \cdots X_N$ where the X_i are i.i.d. Poisson(λ) and N is geometric with success probability r. Standard formulas for random sums tell us

$$ES_N = EXEN = \lambda/r$$
$$\mathrm{var}\,(S_N) = EN \cdot \mathrm{var}\,(X) + (EX)^2 \cdot \mathrm{var}\,(N)$$
$$= \frac{1}{r} \cdot \lambda + \lambda^2 \cdot \frac{1-r}{r^2}$$

To check the form of the two terms, consider the two cases N is constant and X is constant. Using this we have

$$E(S_N(S_N - 1)) = \text{var}(S_N) + (ES_N)^2 - ES_N$$

$$= \frac{1}{r} \cdot \lambda + \lambda^2 \cdot \frac{1-r}{r^2} + \frac{\lambda^2}{r^2} - \frac{\lambda}{r} = \frac{\lambda^2}{r^2}(2-r)$$

so the mean of the size-biased distribution

$$\frac{E(S_N(S_N - 1))}{ES_N} = \lambda \frac{2-r}{r}$$

It follows that the conditions for a giant component are

$$\rho p(2-q)/q > 1 \quad \text{for site percolation}$$
$$\rho p(1+p)/(1-p) > 1 \quad \text{for bond percolation}$$

in agreement with our previous calculations. The main point of this calculation is that using results in Section 3.2 for phase transitions in a graph with a fixed degree distribution, it leads to a rigorous proof. However, I find it comforting that the same critical values emerge from two much different computations.

Critical Exponents. For the rest of their paper, Moore and Newman (2000) are concerned with the values of various critical exponents associated with the percolation process. Their computations are for the Big World graph where cluster growth is a branching process, so they apply equally well to the Erdös–Rényi random graph.

Abstracting the calculation to simplify it, suppose we have a one parameter family of branching processes indexed by their mean μ. If $\mu < 1$ then the total cluster size $\sum_{m=0}^{\infty} Z_m$ has expected value

$$E|\mathcal{C}| = E(\sum_{m=0}^{\infty} Z_m) = \sum_{m=0}^{\infty} \mu^m = \frac{1}{1-\mu}$$

Thus as $\mu \uparrow 1$, $E|\mathcal{C}| \sim (\mu_c - \mu)^{-1}$ and the critical exponent associated with the divergence of the mean cluster size is $\gamma = 1$.

Suppose now that $\mu > 1$ and consider the probability of no percolation ρ which is the solution < 1 of $g(x) = x$. When μ is close to 1, $\rho \approx 1$. Setting $\rho = 1 - a$ and expanding the generating function to second order:

$$g(1-a) = g(1) - ag'(1) + \frac{a^2}{2} g''(b) \quad \text{for some } b \in [1-a, 1]$$

Recalling $g(1) = 1$ and $g'(1) = \mu$, we see that if $g(1-a) = 1-a$ then

$$1 - a = 1 - \mu a + \frac{a^2}{2} g''(b)$$

or $a = 2(\mu - 1)/g''(b)$. As $\mu \downarrow 1$, $g''(b) \rightarrow \mu_2 = \sum_k k(k-1)p_k$, so the critical exponent associated with the survival probability is $\beta = 1$.

Moore and Newman also compute the critical exponent for the asymptotic behavior of the cluster size distribution when $\mu = 1$, but this is the same as the calculation at the end of Section 3.1. The result is

$$P(|\mathcal{C}| = k) \sim bk^{-3/2}$$

To understand this probabilistically, we allow only one individual to reproduce in the branching process at each time, reducing the process to a mean zero random walk in which the time to hit 0 has the same distribution as $|\mathcal{C}|$.

The values we have computed are the "mean-field critical values" which hold for percolation on $\mathbb{Z}^d$ when d is large enough, that is, $d > 6$. Their appearance here indicates that the long-range connections, make the small world very big. Indeed the fact that the diameter grows like $\log n$ compared to $n^{1/d}$ in d-dimensional space, implies that the big and small worlds are essentially infinite dimensional.

What have we just done? Our computations are rigorous results for the branching process. In the random graph, the exponents only appear when we first let $n \rightarrow \infty$ and then let μ approach 1, or set $\mu = 1$ and let $k \rightarrow \infty$. For finite n, the expect values of the average cluster size and the fraction of vertices in the largest component are smooth, and the power law for $P(|\mathcal{C}| = k)$ will have an exponential cutoff for $k = O(n^{2/3})$, see (2.7.4) for the Erdös–Rényi case.

5.4 Ising and Potts Models

The results in this section were inspired by Häggström (2000), but for the details we mostly follow Häggström (1998). In the Potts model, each vertex is assigned a spin $\sigma(x)$ which may take one of q values. Given a finite graph G with vertices V and edges E, for example, the small world, the energy of a configuration is

$$H(\sigma) = 2 \sum_{x,y \in V, x \sim y} 1\{\sigma(x) \neq \sigma(y)\}$$

where $x \sim y$ means x is adjacent to y. Configurations are assigned probabilities proportional to $\exp(-\beta H(\sigma))$ where β is a variable inversely proportional to temperature, and we define a probability measure on $\{1, 2, \ldots q\}^V$ by

$$\nu(\sigma) = Z^{-1} \exp(-\beta H(\sigma))$$

where Z is a normalizing constant that makes the $\nu(\sigma)$ sum to 1. When $q = 2$ this is the Ising model, though in that case it is customary to replace $\{1, 2\}$ by $\{-1, 1\}$, and write the energy as

$$H_2(\sigma) = - \sum_{x,y \in V, x \sim y} \sigma(x)\sigma(y)$$

This leads to the same definition of ν since every pair with $\sigma(x) \neq \sigma(y)$ increases H_2 by 2 from its minimum value in which all the spins are equal, so $H - H_2$ is constant and after normalization the measures are equal.

To study the Potts model on the small world we will use the random–cluster model. This was introduced by Fortuin and Kasteleyn (1972), but Aizenman, Chayes, Chayes, and Newman (1988) were the first use this connection to prove rigorous results. See Grimmett (1995) for a nice survey. The random–cluster model is a $\{0, 1\}$-valued process η on the edges E of the graph:

$$\mu(\eta) = Z^{-1} \left\{ \prod_{e \in E} p^{\eta(e)} (1 - p)^{1 - \eta(e)} \right\} q^{\chi(\eta)}$$

where $\chi(\eta)$ is the number of connected components of η when we interpret 1-bonds as occupied and 0-bonds as vacant, and Z is another normalizing constant. When $q = 1$ this is just a product measure

To relate the two models we introduce the following coupling on $\{1, 2, \ldots, q\}^V \times \{0, 1\}^E$

$$P(\sigma, \eta) = Z^{-1} \left\{ \prod_{e \in E} p^{\eta(e)} (1 - p)^{1 - \eta(e)} \right\} \prod_{x \sim y} 1\{(\sigma(x) - \sigma(y))\eta(x, y) = 0\}$$

In words, if $\eta(x, y) = 1$ then the spins at x and y must agree. It is easy to check, see Theorem 2.1 in Häggström (1998) for detailed proofs of this and the next three results.

Lemma 5.4.1. *If $p = 1 - e^{-2\beta}$ then the projection onto the $\{1, 2, \ldots, q\}^V$ is ν and onto $\{0, 1\}^E$ is μ.*

As a corollary of the coupling we see that

Lemma 5.4.2. *If we pick a random edge configuration according to μ and then assign random values to each connected component of edges the result is ν. Conversely, if we generate $\sigma \in \{1, 2, \ldots, q\}^G$ and then independently assign each edge (x, y) the value 1 with probability p if $\sigma(x) = \sigma(y)$, and probability 0 if $\sigma(x) \neq \sigma(y)$ then the result is μ.*

To begin to analyze the Potts model, we need the following result that follows immediately from the definition.

Lemma 5.4.3. *Fix an edge $e = (x, y)$ and let η^e be the values on $E - e$*

$$\mu(\eta(e) = 1 | \eta^e) = \begin{cases} p & \text{if } x \text{ and } y \text{ are connected in } \eta^e \\ \dfrac{p}{p + q(1 - p)} & \text{otherwise} \end{cases}$$

The next ingredient is a result of Holley (1974), which we consider for the special case of $\{0, 1\}^E$. We say $f : \{0, 1\}^E \to \mathbb{R}$ is increasing if $f(\eta) \le f(\zeta)$ whenever $\eta \le \zeta$, that is, $\eta(e) \le \zeta(e)$ for all $e \in E$. Given two probability measures on $\{0, 1\}^E$, we say that $\mu_1 \le \mu_2$ if $\int f d\mu_1 \le \int f d\mu_2$ for all increasing f.

Lemma 5.4.4. *Let μ_1 and μ_2 be two measures on $\{0, 1\}^E$. Suppose that for every $e \in E$ and every η, ζ with $\eta^e \le \zeta^e$*

$$\mu_1(\eta(e) = 1|\eta^e) \le \mu_2(\zeta(e) = 1|\zeta^e)$$

then $\mu_1 \le \mu_2$.

For a proof see Theorem 3.2 in Häggström (1998).

Introducing the parameters of μ as subscripts, it follows from Lemmas 5.4.3 and 5.4.4 that if $q > 1$

$$\mu_{p,1} \ge \mu_{p,q}$$

and that if $p' \ge p$ and $p'/(p' + q(1 - p')) \ge p$ then

$$\mu_{p,1} \le \mu_{p',q}$$

Theorem 5.4.5. *Let p_c be the critical value for the existence of components of $O(n)$ for percolation on the graph. If*

$$p' > p_I = \frac{qp_c}{1 + (q - 1)p_c} \tag{5.4.1}$$

then $\mu_{p',q}$ also has large components.

Using Lemma 5.4.1 we see that if $\beta > \beta_I$ where $\beta_I = -(1/2)\log(1 - p_I)$ then in $\nu_{\beta,q}$ there is a large clusters of spins all of which have the same value. Turning to concrete examples:

BC small world. The BC small world looks locally like the tree in which each vertex has degree 3. Thinking about the growth of a cluster from a fixed vertex it is easy to see that the critical value for percolation is $p_c = 1/2$. Using Lemma 5.4.5, $p_I = 2/3$. Theorem 5.4.1 gives $\beta_I = (1/2)\log 3 = 0.5493$. It is interesting to note that $1/\beta_I = 1.820$ while simulations of Hong, Kim, and Choi (2002) give $T_c \approx 1.82$. To lead into the next topic which will start to explain this, we begin with some arithmetic that is general:

$$\tanh(\beta_I) = \frac{1 - e^{-2\beta_I}}{1 + e^{-2\beta_I}} = \frac{p_I}{2 - p_I} = p_c$$

since $p_I = 2p_c/(1 + p_c)$.

To explain the significance of this simple calculation, consider the Ising model on a tree with forward branching number $b \ge 2$. b is the degree of vertices -1. Define β_I, the critical value for the Ising model, by $\tanh(\beta_I) = 1/b$. This is the critical value for the onset of "spontaneous magnetization." When $\beta > \beta_I$ if we impose $+1$ boundary conditions at sites a distance n from the root and let $n \to \infty$ then in the resulting limit spins $\sigma(x)$ have positive expected value. When $0 \le \beta \le \beta_I$ there is a unique limiting Gibbs state independent of the boundary conditions, see for example, Preston (1974).

NW small world, $k = 1$. At the end of the 1980s, Russ Lyons, using an idea of Furstenburg, defined a notion of branching number b for trees that are not regular. We will not give the general definition, since for us it is enough that in the case of a tree generated by a multitype branching process, the branching number $b = \lambda$ the growth rate for the process, see Lyons (1990). Lyons (1989) showed that the critical value for percolation $p_c = 1/b$ while if we convert his notation to ours by writing $\beta = J/kT$ where T is the temperature and k is Boltzmann's constant then $\tanh(\beta_I) = 1/b$. See Lyons (2000) for a more recent survey.

The first result gives another derivation of the conclusion $p_c = 1/\lambda$ from the previous section. The second allows us to prove of the upper bound in the next result. The Ising model on small worlds has been studied by physicists, see Barrat and Weigt (2000), Gitterman (2000), Pekalski (2001), and Hong, Kim, and Choi (2002). However, no one seems to have noticed this simple exact result.

Theorem 5.4.6. *For the BC small world or the nearest neighbor NW small world, the critical value for the Ising model has* $\tanh(\beta_I) = p_c$.

Proof. The calculations above show that for $\beta > \beta_I$ there is long-range order in the Ising model in the sense that there are clusters of spins of equal value of size $O(n)$. To prove a result in the other direction we note that if $\beta < \beta_I$ then the Gibbs state on the tree is unique. There is a $c > 0$ so that for most sites x in the graph if we look at the graph in a neighborhood of radius $c \log n$ around x, we see a tree. If we put $+1$'s on the boundary of this tree then what we see inside is larger than the Gibbs state on the small world, but if n is large $P(\sigma(x) = 1) \approx 1/2$. ∎

Spin Glass Transition. Consider the Ising model on the tree with forward branching number b. Define β_c^{SG}, where the superscript SG is for spin glass, by $\tanh \beta_c^{SG} = 1/\sqrt{b}$. The second transition concerns the behavior with free boundary conditions, that is, we truncate the tree at distance n and throw away the sites outside. Bleher, Ruiz, and Zagrebnov (1995) and Ioffe (1996) showed that the limiting state is ergodic when $\beta_c^F < \beta \le \beta_c^{SG}$, but not when $\beta > \beta_c^{SG}$. Here the phrase "spin glass" refers a model on the tree analyzed by Chayes, Chayes, Sethna, and Thouless (1986) and Carlson et al. (1989), which provides the key ideas for the proofs in the two papers previously cited.

To see what this second phase transition means, we consider a model of "Broadcasting on trees" considered by Evans, Kenyon, Peres, and Schulman (2000). Starting at the root, which has some value, say $+1$, each vertex receives the state of its parent with probability $1 - 2\epsilon$ and a randomly chosen state $\in \{-1, 1\}$ with probability

2ϵ. This description is supposed to remind the reader of Lemma 5.4.3, which with Lemma 5.4.1 gives

$$1 - 2\epsilon = \frac{p}{2-p} = \frac{1 - e^{-2\beta}}{1 + e^{-2\beta}} = \tanh(\beta)$$

The four authors show that the probability of correctly reconstructing the spin at the root tends to a limit $> 1/2$ if $1 - 2\epsilon > k^{-1/2}$ and to $1/2$ if $1 - 2\epsilon < k^{-1/2}$.

Question. Does this transition have any meaning for the Ising model on the BC small world?

5.5 Contact Process

Durrett and Jung (2005) have considered the contact process (SIS epidemic) on a multi-dimensional generalization of the BC small world. To make those results fit more easily into the scheme of this chapter, we will for simplicity restrict our attention to the $d = 1$ case. Based on results for the Ising model in the previous section, we should expect that the contact process on the small world should behave like the contact process on a tree, so we begin with an account of those results.

In the contact process on any graph infected sites become healthy at rate 1, and become infected at rate λ times the number of infected neighbors. Let T be a tree in which each vertex has degree $d > 2$ and let 0 be a distinguished vertex (the origin) of the tree. Let A_t^0 be the set of infected sites at time t on the tree starting from 0 occupied. We define two critical values:

$$\lambda_1 = \inf\{\lambda : \mathbb{P}(|A_t^0| = 0 \text{ eventually}) < 1\} \qquad (5.5.1)$$
$$\lambda_2 = \inf\{\lambda : \liminf_{t\to\infty} \mathbb{P}(0 \in A_t^0) > 0\}.$$

We call λ_1 the weak survival critical value and λ_2 the strong survival critical value. Pemantle (1992) showed that for trees with $d \geq 4$, $\lambda_1 < \lambda_2$. Pemantle (1992) and Liggett (1996) who extended the result to trees with $d = 3$, did this by finding numerical bounds on the two critical values which showed they were different. Later, Stacey (1996) found a proof that did not rely on numerical bounds.

To explain the reason for the two critical values, consider branching random walk, which is a contact process without the restriction of one particle per site. In this process each particle dies at rate 1, and for each neighbor gives birth at rate λ to a new particle at that site. If Z_t^0 is the number of particles at time t starting from a single particle at time 0 then $E Z_t^0 = e^{(\lambda d - 1)t}$ so $\lambda_1 = 1/d$. If we let S_t be a random walk on the tree that jumps at rate λd to a randomly chosen neighbor then the expected number of particles at 0 at time t has

$$E Z_t^0(0) = e^{(\lambda d - 1)t} P(S_t = 0)$$

For a detailed proof of a similar fact see (4.4.1).

Since the distance from the origin on the tree is a random walk that steps $+1$ with probability $(d-1)/d$ and -1 with probability $1/d$ (except at 0 where all steps are $+1$) it is not hard to show that

$$(1/t)\log P(S_t = 0) \to -\rho_d < 0$$

so $\lambda_2 d - 1 - \rho_d = 0$ or $\lambda_2 = (1 + \rho_d)/d$. For more on the two phase transitions in branching random walk see Madras and Schinazi (1992).

Our version of the BC small world, which we will call BC_m, will be as follows. We start with a ring $\mathbb{Z} \bmod L$ and connect each vertex to all other vertices within distance m. We require L to be even so that we can partition the L vertices into $L/2$ pairs. Consider all such partitions and then pick one at random. A new edge is then drawn between each pair of vertices in the chosen partition. The reason for insisting that all individuals have exactly one long-range neighbor is that we can define an associated "big world" graph $\mathcal{B}_m$ that is nonrandom. Algebraically, $\mathcal{B}_m$ consists of all vectors $\pm(z_1, \ldots, z_n)$ with $n \geq 1$ components with $z_j \in \mathbb{Z}$ and $z_j \neq 0$ for $j < n$. Neighbors in the positive half-space are defined as follows: a point $+(z_1, \ldots, z_n)$ is adjacent to $+(z_1, \ldots, z_n + y)$ for all y with $0 < |y| \leq m$ (these are the short-range neighbors of $+(z_1, \ldots, z_n)$). The long-range neighbor is

$$
\begin{aligned}
&+(z_1, \ldots, z_n, 0) && \text{if } z_n \neq 0\\
&+(z_1, \ldots, z_{n-1}) && \text{if } z_n = 0, n > 1\\
&-(0) && \text{if } z_n = 0, n = 1.
\end{aligned}
$$

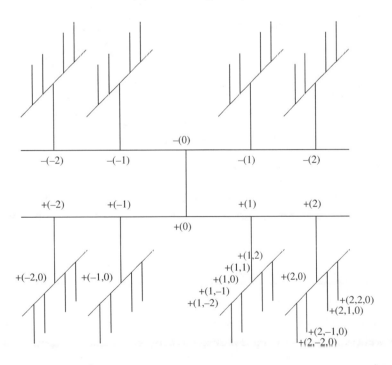

We will consider the discrete time contact process. On either the small world or the big world, an infected individual lives for one unit of time. During its infected period it will infect some of its neighbors. All infection events are independent, and each site that receives at least one infection is occupied with an infected individual at the next time. A site infects itself or its short-range neighbors with probability $\alpha/(2m + 1)$. It infects its long-range neighbor with probability β. To have a one parameter family of models we think of fixing $r = \alpha/\beta$ and varying $\lambda = \alpha + \beta$.

We will use B_t to denote the contact process on the big world and ξ_t for the contact process on the small world. It is easy to see that if $\alpha + \beta < 1$ the infection on the big world will die out. Our first result shows that this trivial necessary condition becomes exact when the range m is large.

Theorem 5.5.1. *If $\alpha + \beta > 1$ then the contact process on the big world survives for large m.*

The proofs of this and the other results are somewhat lengthy, so we refer the reader to Durrett and Jung (2005) for details.

To obtain a lower bound on λ_2, we use the fact that strong survival of the contact process on $\mathcal{B}_m$ implies strong survival of the branching random walk on $\mathcal{B}_m$. Let $\lambda_2^{brw}(m)$ be the strong survival critical value of the branching random walk. To compute the limit of $\lambda_2^{brw}(m)$, we define the "comb" of degree m, $\mathcal{C}_m$, by restricting $\mathcal{B}_m$ to vertices of the form $\{+(z), +(z, 0), -(0)\}$ and all edges between any of these vertices. As before, $+(z)$ and $+(z, 0)$ are long-range neighbors as are $+(0)$ and $-(0)$. The short-range neighbors of $+(z)$ are $+(z + y)$ for $0 < |y| \leq m$. The vertices $+(z, 0)$ and $-(0)$ have no short-range neighbors. To see the reason for the name look at the picture.

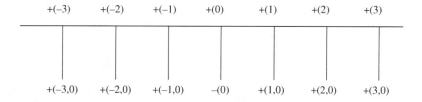

Viewing particles on the top row as type 1 and those on the bottom row as type 2, we have a two type branching process with mean matrix:

$$\begin{pmatrix} \alpha & \beta \\ \beta & 0 \end{pmatrix}$$

Results for multitype branching processes imply that the branching random walk on the comb survives if the largest eigenvalue of the matrix is larger than 1. Solving the quadratic equation $(\alpha - \lambda)(-\lambda) - \beta^2 = 0$ the largest root is

$$\frac{\alpha + \sqrt{\alpha^2 + 4\beta^2}}{2}$$

A little algebra shows that this is larger than 1 exactly when $\alpha^2 + 4\beta^2 > (2 - \alpha)^2$ or $\alpha + \beta^2 > 1$. This is an upper bound on the strong survival critical value of the branching random walk when what we need is a lower bound but it motivates the following.

Theorem 5.5.2. *If $\alpha + \beta^2 < 1$ then there is no strong survival in the contact process on the big world for large m.*

Comparing the above with Theorem 5.5.1 shows that for any $r = \alpha/\beta$, $\lambda_1 < \lambda_2$ for large m. When $m = 1$ and $\alpha = \beta$ the big world is a tree of degree 3 and we have $\lambda_1 < \lambda_2$ in that case as well. It is reasonable to conjecture that for any range m and ratio r we have $\lambda_1 < \lambda_2$ but this seems difficult to prove.

Since the small world is a finite graph, the infection will eventually die out. However, by analogy with results for the d-dimensional contact process on a finite set, we expect that if the process does not become extinct quickly, it will survive for a long time. Durrett and Liu (1988) showed that the supercritical contact process on $[0, L)$ survives for an amount of time of order $\exp(cL)$ starting from all ones, while Mountford (1999) showed that the supercritical contact process on $[0, L)^d$ survives for an amount of time of order $\exp(cL^d)$. At the moment we are only able to prove the last conclusion for the following modification of the small world contact process: each infected site infects its short-range neighbors with probability $\alpha/(2m + 1)$ and its long-range neighbor with probability β, but now in addition, it infects a random neighbor (chosen uniformly from the grid) with probability $\gamma > 0$.

From a modeling point of view, this mechanism is reasonable. In addition to long-range connections with friends at school or work, one has random encounters with people one sits next to on airplanes or meets while shopping in stores. In the language of physics, the model with $\gamma = 0$ has a quenched (i.e., fixed) random environment, while the model with $\beta = 0$ has an annealed environment.

Our strategy for establishing prolonged survival is to show that if the number of infected sites drops below ηL, it will with high probability rise to $2\eta L$ before dying out. To do this we use the random connections to spread the particles out so that they can grow independently. Ideally, we would use the long-range connections (instead of the random connections) to achieve this; however, we have to deal with unlikely but annoying scenarios such as all infected individuals being long-range neighbors of sites that are respectively short-range neighbors of each other.

Theorem 5.5.3. *Consider the modified small world model on $\mathbb{Z}$ mod L with random infections at rate $\gamma > 0$. If $\lambda > \lambda_1$ and we start with all infected individuals then*

there is a constant $c > 0$ so that the probability the infection persists to time $\exp(cL)$
tends to $\bar{1}$ *as* $L \to \infty$.

This result shows that prolonged persistence occurs for $\lambda > \lambda_1$. The next describes a change in the qualitative behavior that occurs in the contact process at λ_2. Let $\tau_B = \min\{t : B_t^0 = \emptyset\}$ be the extinction time of the contact process on the big world. Let $\sigma_B = \min\{t : B_t^0 = \emptyset \text{ or } 0 \in B_t^0\}$ be the first time that the infection either dies out or comes back to the origin starting from one infection there at time 0. Let $\tau_S = \min\{t : \xi_t^0 = \emptyset\}$ and $\sigma_S = \min\{t \geq 1 : \xi_t^0 = \emptyset \text{ or } 0 \in \xi_t^0\}$ be the corresponding times for the contact process on the small world.

Theorem 5.5.4. *Writing $\Rightarrow$ for convergence in distribution as $L \to \infty$ we have*
(a) τ_S is stochastically bounded above by τ_B and $\tau_S \Rightarrow \tau_B$
(b) σ_S is stochastically bounded above by σ_B and $\sigma_S \Rightarrow \sigma_B$.

Intuitively, when $\lambda_1 < \lambda < \lambda_2$, the infection cannot spread without the help of the long range so even if the infection starts at 0 and does not die out globally then it dies out locally and takes a long time to return to 0.

Open problem. Consider the Ising model on BC_m. Is $\tanh(\beta_I) = p_c$ the critical value for percolation on the big world?

6

Random Walks

In this chapter we will study the mixing times of random walks on our random graphs. In order to make our treatment self-contained, we will give an account of the Markov chain results we use.

6.1 Spectral Gap

Consider a Markov chain transition kernel $K(i, j)$ on $\{1, 2, \ldots n\}$ with reversible stationary distribution π_i, that is, $\pi_i K(i, j) = \pi_j K(j, i)$. To measure convergence to equilibrium we will use the *relative pointwise distance*

$$\Delta(t) = \max_{i,j} \left| \frac{K^t(i, j)}{\pi_j} - 1 \right|$$

which is larger than the total variation distance

$$\Delta(t) \geq \max_i \sum_j \left| \frac{K^t(i, j)}{\pi_j} - 1 \right| \pi_j = \max_i \sum_j |K^t(i, j) - \pi_j|$$

Let D be a diagonal matrix with entries $\pi_1, \pi_2, \ldots \pi_n$ and $a = D^{1/2} K D^{-1/2}$. Since

$$a(i, j) = \pi_i^{1/2} K(i, j) \pi_j^{-1/2} = \pi_j^{1/2} K(j, i) \pi_i^{-1/2} = a(j, i)$$

matrix theory tells us that $a(i, j)$ has real eigenvalues $1 = \lambda_0 \geq \lambda_1 \geq \cdots \lambda_{n-1} \geq -1$. Let $\lambda_{\max} = \max\{\lambda_1, |\lambda_{n-1}|\}$ be the largest in magnitude of $\lambda_1, \ldots, \lambda_{n-1}$. The next result is from Sinclair and Jerrum (1989), but similar results can be found in many other places.

Theorem 6.1.1. *Let k be the transition matrix of an irreducible reversible Markov chain on $\{1, 2, \ldots n\}$ with stationary distribution π and let $\pi_{\min} = \min_j \pi_j$. Then*

$$\Delta(t) \leq \frac{\lambda_{\max}^t}{\pi_{\min}}$$

153

Proof. Since a is symmetric, we can select an orthonormal basis $e_m, 0 \leq m < n$ of eigenvectors of a, and a has spectral decomposition:

$$a = \sum_{m=0}^{n-1} \lambda_m e_m e_m^T$$

The matrix $B_m = e_m e_m^T$ has $B_m^2 = B_m$, and $B_\ell B_m = 0$ if $\ell \neq m$ so

$$a^t(i, j) = \sum_{m=0}^{n-1} \lambda_m^t B_m(i, j) = \sum_{m=0}^{n-1} \lambda_m^t e_m(i) e_m(j)$$

$e_0(i) = \pi_i^{1/2}$ so

$$K^t(i, j) = (D^{-1/2} a^t D^{1/2})_{i,j} = \pi_j + \sqrt{\frac{\pi_j}{\pi_i}} \sum_{m=1}^{n-1} \lambda_m^t e_m(i) e_m(j)$$

From this it follows that

$$\Delta(t) = \max_{i,j} \frac{\left| \sum_{m=1}^{n-1} \lambda_m^t e_m(i) e_m(j) \right|}{\sqrt{\pi_i \pi_j}} \leq \lambda_{\max}^t \frac{\max_{i,j} \sum_{m=1}^{n-1} |e_m(i)||e_m(j)|}{\pi_{\min}}$$

The Cauchy–Schwarz inequality implies

$$\sum_{m=1}^{n-1} |e_m(i)||e_m(j)| \leq \left(\sum_{m=1}^{n-1} |e_m(i)|^2 \sum_{m=1}^{n-1} |e_m(j)|^2 \right)^{1/2}$$

To see that $\sum_{m=1}^{n-1} |e_m(i)|^2 \leq 1$ note that if δ_i is the vector with 1 in the ith place and 0 otherwise then expanding in the orthonormal basis $\delta_i = \sum_{m=0}^{n-1} e_m(i) e_m$, so the desired result follows by taking the L^2 norm of both sides of the equation. ∎

Given a reversible Markov transition kernel $K(x, y)$ we define the *Dirichlet form* by

$$\mathcal{E}(f, g) = \frac{1}{2} \sum_{x,y} (f(x) - f(y))(g(x) - g(y))\pi(x)K(x, y) \qquad (6.1.1)$$

Introducing the inner product $< f, g >_\pi = \sum_x f(x)g(x)\pi(x)$, a little algebra shows

$$\mathcal{E}(f, f) = < f, (I - K)f >_\pi$$

If we define the variance by $\mathrm{var}_\pi(f) = E_\pi(f - E_\pi f)^2$ then the spectral gap can be computed from the variational formula

$$1 - \lambda_1 = \min\{\mathcal{E}(f, f) : \mathrm{var}_\pi(f) = 1\} \qquad (6.1.2)$$

To see this note that $\mathcal{E}(f, f)$ is not affected by subtracting a constant from f so

$$1 - \lambda_1 = \min\{< f, f >_\pi - < f, Kf >_\pi : E_\pi f = 0, < f, f >_\pi = 1\}$$

and the result follows from the usual variational formula for λ_1 for the nonnegative symmetric matrix $a_{i,j} = \pi(i)K(i,j)$, that is,

$$\lambda_1 = \max \left\{ \sum_i x_i a_{i,j} x_j : \sum_i x_i^2 = 1 \right\}$$

For some of our results we will consider continuous time chains. If jumps occur at rate one then there are a Poisson mean t jumps by time t so the transition probability is

$$H_t(x, y) = e^{-t} \sum_{m=0}^{\infty} \frac{t^m}{m!} K^m(x, y)$$

If λ_i is an eigenvalue of K then $e^{-t(1-\lambda_i)}$ is an eigenvalue of H_t. Thus there are no negative eigenvalues to worry about and we have

Theorem 6.1.2. *If $\beta = 1 - \lambda_1$ is the spectral gap of K then*

$$\max_{x,y} \left| \frac{H_t(x, y)}{\pi(y)} - 1 \right| \le \frac{e^{-\beta t}}{\pi_{\min}}$$

Proof. Let $H_t g(x) = \sum_y H_t(x, y)g(y)$. Differentiating the series for H_t we have

$$\frac{\partial}{\partial t} H_t g = -H_t g + e^{-t} \sum_{m=1}^{\infty} \frac{t^{m-1}}{(m-1)!} \sum_{z,y} K(x, z) K^{m-1}(z, y)g(y) = -(I - K)H_t g$$

and it follows that

$$\frac{\partial}{\partial t} \sum_x \pi(x)(H_t g(x))^2 = -2 < H_t g, (I - K)H_t g >_\pi = -2\mathcal{E}(H_t g, H_t g)$$

Applying the last result to $g(x) = f(x) - \pi(f)$ where $\pi(f) = E_n f$ and letting $u(t) = \sum_x \pi(x)(H_t g(x))^2 = \|H_t g\|_2^2$ we have $u'(t) = -2\mathcal{E}(H_t g, H_t g) \le -2\beta u(t)$ by the variational characterization of the spectral gap. Integrating the differential inequality

$$\|H_t f - \pi(f)\|_2^2 = u(t) \le e^{-2\beta t} u(0) = e^{-2\beta t} \operatorname{var}_\pi f \qquad (6.1.3)$$

where $\operatorname{var}_\pi f = \|f - \pi(f)\|_2^2$ is the variance of f under π.

Define the dual transition probability by

$$\hat{H}_t(x, y) = \frac{\pi(y)H_t(y, x)}{\pi(x)}$$

For a probabilist this is the time-reversed chain:

$$\hat{H}_t(x, y) = \frac{P_\pi(X_0 = y, X_t = x)}{P_\pi(X_t = x)} = P_\pi(X_0 = y | X_t = x)$$

For analysts this is the adjoint operator

$$< f, H_t g >_\pi = \sum_{x,y} f(y)\pi(y)H_t(y, x)g(x)$$

$$= \sum_{x,y} \pi(x)\hat{H}_t(x, y)f(y)g(x) =< \hat{H}_t f, g_\pi >$$

Letting $h_t^x(y) = H_t(x, y)/\pi(y)$ and

$$f_x(z) = \begin{cases} 1/\pi(x) & z = x \\ 0 & \text{otherwise} \end{cases}$$

we have $\pi(f_x) = 1$ and

$$\hat{H}_t f_x(y) = \sum_z \hat{H}_t(y, z)f_x(z) = \frac{\hat{H}_t(y, x)}{\pi(x)} = \frac{H_t(x, y)}{\pi(y)} = h_t^x(y)$$

Since $\hat{H}_t$ is reversible with respect to π, (6.1.3) implies

$$\|h_t^x - 1\|_2^2 \le e^{-2\beta t} \frac{1 - \pi(x)}{\pi(x)} \tag{6.1.4}$$

Using the Markov property, adding and subtracting 1, and using reversibility

$$\left| \frac{H_t(x, y)}{\pi(y)} - 1 \right| = \left| \sum_z \left[\frac{H_{t/2}(x, z)H_{t/2}(z, y)}{\pi(z)\pi(y)} - 1 \right] \pi(z) \right|$$

$$= \left| \sum_z \left(\frac{H_{t/2}(x, z)}{\pi(z)} - 1 \right) \left(\frac{H_{t/2}(z, y)}{\pi(y)} - 1 \right) \pi(z) \right|$$

$$= \left| \sum_z \left(\frac{H_{t/2}(x, z)}{\pi(z)} - 1 \right) \left(\frac{H_{t/2}(y, z)}{\pi(z)} - 1 \right) \pi(z) \right|$$

Using the Cauchy–Schwarz inequality now and (6.1.4) the above

$$\le \|h_{t/2}^x - 1\|_2 \cdot \|h_{t/2}^y - 1\|_2 \le e^{-\beta t} \frac{1}{\sqrt{\pi(x)\pi(y)}}$$

from which the desired result follows.   ∎

6.2 Conductance

Suppose for the moment that we have a general reversible transition probability, write $Q(x, y) = \pi(x)K(x, y)$, and define

$$h = \min_{\pi(S) \le 1/2} \frac{Q(S, S^c)}{\pi(S)}$$

where $Q(S, S^c) = \sum_{x \in S, y \in S^c} Q(x, y)$. Since this is the size of the boundary of S when edge (x, y) is assigned weight $Q(x, y)$, we will sometimes write this as $|\partial S|$. Our next result is Lemma 3.3.7 in Saloff-Coste (1996). His $I = 2h$ so the constants are different. Saloff-Coste attributes the result to Diaconis and Stroock (1991), who in turn named the result Cheeger's inequality in honor of the eigenvalue bound in differential geometry.

Theorem 6.2.1. *The spectral gap has*

$$\frac{h^2}{2} \le 1 - \lambda_1 \le 2h$$

Proof. Taking $f = 1_S$ in the variational formula (6.1.2) we have

$$\mathcal{E}(1_S, 1_S) = Q(S, S^c)$$

and $\text{var}_\pi(1_S) = \pi(S)(1 - \pi(S))$, so $1 - \lambda_1 \le Q(S, S^c)/\pi(S)(1 - \pi(S))$. The right-hand side is the same for S and S^c, so we can restrict our attention to $\pi(S) \le 1/2$. Since $1 - \pi(S) \ge 1/2$, we have $1 - \lambda_1 \le 2h$.

For the other direction, let $F_t = \{x : f(x) \ge t\}$ and let f_t be the indicator function of the set F_t. Since only differences $f(x) - f(y)$ appear in $\mathcal{E}(f, f)$, defined in (6.1.1), we can without loss of generality suppose that the median of f is 0, that is, $\pi(F_t) \le 1/2$ for $t > 0$, and $\pi(F_t^c) \le 1/2$ for $t < 0$. Our next step is to compute something that would be the Dirichlet form if we had squared the increment.

$$\frac{1}{2} \sum_{x,y} |f(x) - f(y)| Q(x, y) = \sum_{f(x) > f(y)} (f(x) - f(y)) Q(x, y)$$

$$= \sum_{x,y} \int_{-\infty}^\infty 1_{\{f(y) < t < f(x)\}} Q(x, y) \, dt$$

$$= \int_0^\infty |\partial F_t| \, dt + \int_{-\infty}^0 |\partial F_t^c| \, dt$$

$$\ge h \left(\int_0^\infty \pi(F_t) \, dt + \int_{-\infty}^0 \pi(F_t^c) \, dt \right) = h\pi(|f|)$$

Continuing to suppose that the median of f is 0, let $g = f^2 \text{sgn}(f)$, where $\text{sgn}(x) = 1$ if $x > 0$, $\text{sgn}(x) = -1$ if $x < 0$, and $\text{sgn}(0) = 0$. $|g| = f^2$ so the last inequality implies

$$2h\pi(f^2) \le \sum_{x,y} |g(x) - g(y)| Q(x, y) \le \sum_{x,y} |f(x) - f(y)|(|f(x)| + |f(y)|) Q(x, y)$$

To check the last inequality, we can suppose without loss of generality that $f(x) > 0$ and $f(x) > f(y)$. If $f(y) > 0$ we have an equality, while if $f(y) < 0$ we have

$f^2(x) + f^2(y) < (|f(x)| + |f(y)|)^2$. Using the Cauchy–Schwarz inequality now the above is

$$\leq \left(\sum_{x,y} (f(x) - f(y))^2 Q(x, y) \right)^{1/2} \cdot \left(\sum_{x,y} (|f(x)| + |f(y)|)^2 Q(x, y) \right)^{1/2}$$

$$\leq (2\mathcal{E}(f, f))^{1/2} (4\pi(f^2))^{1/2}$$

Rearranging gives $(2\mathcal{E}(f, f))^{1/2} \geq h(\pi(f^2))^{1/2}$. Squaring we have

$$\mathcal{E}(f, f) \geq \frac{h^2}{2} \pi(f^2) \geq \frac{h^2}{2} E_\pi (f - E_\pi f)^2$$

which proves the desired result. ∎

Let G be a finite connected graph, $d(x)$ be the degree of x, and write $x \sim y$ if x and y are neighbors. We can define a transition kernel by $K(x, x) = 1/2$, $K(x, y) = 1/2d(x)$ if $x \sim y$ and $K(x, y) = 0$ otherwise. The 1/2 probability of staying put means that we don't have to worry about periodicity or negative eigenvalues. Our K can be written $(I + p)/2$ where p is another transition probability, so all of the eigenvalues of K are in $[0, 1]$, and $\lambda_{\max} = \lambda_1$.

$\pi(x) = d(x)/D$ where $D = \sum_{y \in G} d(y)$, defines a reversible stationary distribution since $\pi(x)K(x, y) = 1/2D = \pi(y)K(y, x)$. Letting $e(S, S^c)$ is the number of edges between S and S^c, and $\text{vol}(S)$ be the sum of the degrees in S, we have

$$h = \frac{1}{2} \min_{\pi(S) \leq 1/2} \frac{e(S, S^c)}{\text{vol}(S)}$$

When $d(x) \equiv d$, $h = \iota/2d$ where

$$\iota = \min_{|S| \leq n/2} \frac{e(S, S^c)}{|S|}$$

is the *edge isoperimetric constant*.

To illustrate the use of Theorem 6.2.1 and to show that one cannot get rid of the power 2 from the lower bound, consider random walk on the circle $\mathbb{Z}$ mod n in which we stay put with probability 1/2 and jump from x to $x \pm 1$ with probability 1/4 each. Taking $S = \{1, 2, \ldots n/2\}$ we see that

$$\iota = \frac{2}{n/2} = 4/n$$

To bound the spectral gap, we let $f(x) = \sin(\pi x/n)$. Since $\sin(a + b) = \sin a \cos b + \sin b \cos a$ we have

$$(I - K)f(x) = f(x)(1 - \cos(\pi/n))/2$$

and $1 - \lambda_1 \leq (1 - \cos(\pi/n))/2 \sim \pi^2/4n^2$ as $n \to \infty$. Using Theorem 6.1.1 gives an upper bound on the convergence time of order $O(n^2 \log n)$. However, using the local central limit theorem for random walk on $\mathbb{Z}$ it is easy to see that $\Delta(t) \le \epsilon$ at a time $K_\epsilon n^2$.

Mixing Times and the Conductance Profile. Since we are lazy, we will suppose in what follows that the chain is as well: that is, $K(x, x) \ge 1/2$ for all x, and we will not give the proofs of these more sophisticated results. Given an initial state i, it is possible to define stopping times T so that X_T has the stationary distribution. Define $H(i, \pi)$ the minimum value of $E_i T$ for all such stopping times and let $\mathcal{H} = \max_i H(i, \pi)$. Define the mixing time

$$T_{\text{mix}} = \max_i \min\{t : d_{TV}(K^t(i, \cdot), \pi) < 1/e\}$$

The cutoff $1/e$ is somewhat arbitrary. The important thing is that it is small enough to allow us to conclude that if $t \ge \ell T_{\text{mix}}$ then $d_{TV}(K^t(i, \cdot), \pi) < (2/e)^\ell$. Aldous (1982) has shown, see also Aldous, Lovász, and Winkler (1997), that $C_1 \mathcal{H} \le T_{\text{mix}} \le C_2 \mathcal{H}$. Define the conductance profile by

$$\psi(x) = \min_{S:0<\pi(s)\le x} \frac{Q(S, S^c)}{\pi(S)\pi(S^c)}$$

Lovász and Kannan (1999) have shown that

$$\mathcal{H} \le 32 \int_{\pi_{\min}}^{1/2} \frac{dx}{x\psi(x)^2}$$

Morris and Peres (2003) used their notion of evolving sets to sharpen this result to

$$\text{If} \quad n \ge \int_{\pi(i)\wedge\pi(j)}^{4/\epsilon} \frac{4\,dx}{x\psi(x)^2} \quad \text{then} \quad \left| \frac{K^n(i, j)}{\pi(j)} - 1 \right| \le \epsilon$$

These results are useful for improving rate of convergence results in some examples. However, in some of our favorite examples the worst conductance occurs for small sets, so we will instead use a recent result of Fountoulakis and Reed (2007).

Theorem 6.2.2. *If $\psi(x)$ be the minimum $Q(S, S^c)/\pi(S)\pi(S^c)$ over all connected sets S with $x/2 \le \pi(S) \le x$ then*

$$T_{\text{mix}} \le C \int_{\pi_{\min}}^{1/2} \frac{dx}{x\psi(x)^2}$$

6.3 Fixed Degree Distribution

We begin by considering the random r-regular graph and define a random walk that stays put with probability $1/2$ and jumps to each of its r neighbors with probability $1/2r$. To be precise in the face of parallel edges and self-loops, we pick one of the

r edges incident to the current vertex and jump to the vertex at the other end. Since vertices have constant degree, the uniform distribution $\pi(x) = 1/n$ is stationary. As explained in Section 6.2, we can bound the convergence time by bounding the isoperimetric number, ι. For this we can use a very precise result due to Bollobás (1988).

Theorem 6.3.1. *Let $r \geq 3$ and $0 < \eta < 1$ be such that*

$$2^{4/r} < (1-\eta)^{1-\eta}(1+\eta)^{1+\eta}$$

Then asymptotically almost surely a random r-regular graph has $\iota \geq (1-\eta)r/2$.

At first glance, it is hard to see for what values of η the right-hand side is bigger than the left. For $0 < h < 1$,

$$\frac{d}{dh}\left[(1-h)\log(1-h) + (1+h)\log(1+h)\right] = \log\left(\frac{1+h}{1-h}\right) > 0$$

Using a calculator or computer, one can see that when $r = 3$ we need to take $\eta = 0.878$ and when $r = 10$, $\eta = 0.514$. Expanding the logarithms to first order we see that if r is large $\eta^2 \approx (2\log 2)/r \to 0$. To see that the constant cannot be better than $r/2$, let $S = \{1, 2, \ldots n/2\}$, and note that when the points in S pick their r neighbors, the probability of picking a point in S^c is $1/2$. The law of large numbers implies that $e(S, S^c) \approx rn/2$. Of course, some sets of size $n/2$ will not have this typical behavior. Estimating the large deviations leads to a constant $(1-\eta)r/2$.

Here we will prove a more general result with a worse constant.

Theorem 6.3.2. *Consider a random graph with a fixed degree distribution in which the minimum degree is $r \geq 3$. There is a constant $\alpha_0 > 0$ so that $h \geq \alpha_0$.*

From this it follows that the mixing time is $O(\log n)$. The condition $r \geq 3$ is necessary since if there is a positive density of vertices of degree 2 then there will be paths of length $O(\log n)$ in which each vertex has degree 2 and if we start in the middle of the path then the mixing time will be $\geq O(\log^2 n)$.

Proof of Theorem 6.3.2. We begin our computation by considering only the random r-regular graph. Let $f(m)$ be the number of ways of dividing m objects into pairs.

$$f(m) = \frac{m!}{(m/2)!2^{m/2}} \tag{6.3.1}$$

Let $P(u, s)$ be the probability that there is a set U with $|U| = u$ and $e(U, U^c) = s$.

$$P(u, s) \leq \binom{n}{u}\binom{ru}{s}\binom{r(n-u)}{s}s!\,f(ru-s)f(r(n-u)-s)\frac{1}{f(rn)} \tag{6.3.2}$$

To see this recall that in the random configuration model we make r copies of each vertex and then pair the rn mini-vertices at random, which can be done in $f(rn)$ ways. We can choose the set U in $\binom{n}{u}$ ways. We can pick the left ends of the edges to connect U to U^c in $\binom{ru}{s}$ ways, the right ends in $\binom{r(n-u)}{s}$ ways, and then pair them in $s!$ ways. There are $(ru - s)/2$ edges that connect U to U and these can be attached in $f(ru - s)$ ways. There are $(r(n - u) - s)/2$ edges that connect U^c to U^c and these can be attached in $f(r(n - u) - s)$ ways.

Let $s(u)$ be the largest integer less than $(1 - \eta)ru/2$. Bollobás "proves" his result by asserting that if $0 \leq s \leq s' \leq s(u)$ then $P(u, s) \leq P(u, s')$, so it is enough to show that $P(u, s(u)) = o(n^{-2})$. He then claims it is enough to prove the result for $u = n/2$ and in that case "straightforward, though tedious calculations" give the result. Not being able to fill in the details, I turned to Gkantsidis, Mihail, and Saberi (2003) for help. To make it easier to compare with their argument we change values $D = rn$, $k = ru$, and $s = \alpha k$. This converts (6.3.2) into

$$\binom{D/r}{k/r}\binom{k}{\alpha k}\binom{D - k}{\alpha k}\frac{(\alpha k)! f(k - \alpha k) f(D - k - \alpha k)}{f(D)} \tag{6.3.3}$$

Their formula (10) is this with $s! = (\alpha k)!$ replaced by the larger $f(2\alpha k)$. They also have a factor αk to account for $1 \leq s \leq \alpha k$.

Before we start to work on (6.3.3) we observe that this formula is valid for a graph with fixed degree distribution and minimum degree r. In this case D is the sum of the degrees in the graph and k is the volume, that is, the number of mini-vertices. This gives an upper bound on the probability since the number of sets of vertices with volume k is at most $\binom{D/r}{k/r}$, the maximum occurring when all of the vertices have degree r.

To bound the binomial coefficients, the following lemma is useful.

Lemma 6.3.3.

$$\binom{n}{m} \leq \frac{n^m}{m!} \leq \frac{n^m}{m^m e^{-m}}$$

Proof. The first inequality follows from $n(n - 1) \cdots (n - m + 1) \leq n^m$. For the second we note that the series expansion of e^m has only positive terms so $e^m > m^m/m!$. ∎

From Lemma 6.3.3, we see that the three binomial coefficients in (6.3.3) are

$$\leq \left(\frac{De}{k}\right)^{k/r}\left(\frac{e}{\alpha}\right)^{2\alpha k}\left(\frac{D - k}{k}\right)^{\alpha k} \tag{6.3.4}$$

Here, to prepare for a later step, we have transferred part of the bound for the third term into the second.

To bound the f's in (6.3.3) we use Stirling's formula to conclude

$$f(m) = \frac{m!}{(m/2)!2^{m/2}} \sim C\frac{m^{m+1/2}e^{-m}}{(m/2)^{m/2+1/2}e^{-m/2}2^{m/2}} = C(m/e)^{m/2}$$

From this we see that the fraction in (6.3.3) is

$$\leq Ck^{1/2}\frac{(\alpha k/e)^{\alpha k}(k(1-\alpha)/e)^{k(1-\alpha)/2}((D-(1+\alpha)k)/e)^{(D-(1+\alpha)k)/2}}{(D/e)^{D/2}}$$

$$= Ck^{1/2}(\alpha k)^{\alpha k}D^{-\alpha k}\left(\frac{k(1-\alpha)}{D}\right)^{k(1-\alpha)/2}\left(1-\frac{(1+\alpha)k}{D}\right)^{(D-(1+\alpha)k)/2} \quad (6.3.5)$$

since the exponents in the numerator sum to $D/2 - \alpha$.

Combining (6.3.4) and (6.3.5) gives an upper bound

$$\leq Ck^{1/2}\left(\frac{De}{k}\right)^{k/r}\left(\frac{e^2}{\alpha}\right)^{\alpha k}\left(\frac{D-k}{D}\right)^{\alpha k}$$

$$\cdot\left(\frac{k(1-\alpha)}{D}\right)^{k(1-\alpha)/2}\left(1-\frac{(1+\alpha)k}{D}\right)^{(D-(1+\alpha)k)/2}$$

Ignoring the $Ck^{1/2}$'s, the first term is the first term from (6.3.4), the second and third terms come from combining the second and third terms of (6.3.4) and with the first and second terms of (6.3.5), while the remainder of the formula comes from (6.3.5). Using $\alpha > 0$ and $D - k < D$ and rearranging we have

$$= Ck^{1/2}e^{k/r}\left(\frac{e^2}{\alpha}\right)^{\alpha k}\cdot\left(\frac{k}{D}\right)^{k(1-\alpha)/2-k/r}\left(1-\frac{(1+\alpha)k}{D}\right)^{(D-(1+\alpha)k)/2}$$

Setting $\beta = e^2/\alpha$ and $\gamma = (1-\alpha)/2 - 1/r$ we have

$$\leq Ck^{1/2}e^{k/r}\beta^{\alpha k}\cdot\left(\frac{k}{D}\right)^{\gamma k}\left(1-\frac{(1+\alpha)k}{D}\right)^{(D-(1+\alpha)k)/2} \quad (6.3.6)$$

Comparing with formula (17) in Gkantsidis, Mihail, and Saberi (2003), we see that apart from the differences that result from our use of $(\alpha k)!$ instead of $f(2\alpha k)$, they are missing the $e^{k/r}$ and we have retained an extra term to compensate for the error.

Let

$$G(k) = e^{k/r}\beta^{\alpha k}\cdot\left(\frac{k}{D}\right)^{\gamma k}\left(1-\frac{(1+\alpha)k}{D}\right)^{(D-(1+\alpha)k)/2}$$

$Ck^{1/2} \leq Cn^{1/2}$ so we can show $h \geq \alpha_0$ by showing that for $0 \leq \alpha \leq \alpha_0$

$$\sup_{1\leq k\leq D/2} G(k) = o(n^{-5/2})$$

because then we can sum our estimate over $k \leq D/2$ and $s = \alpha k$ with $\alpha \leq \alpha_0$ and end up with a result that is $o(1)$.

$\beta^{\alpha k} = \exp(\eta k)$ where $\eta = \alpha \log(e^2/\alpha) \to 0$ as $\alpha \to 0$. Ignoring this term, and setting $k = D/2$, $\alpha = 0$

$$G(D/2) = e^{D/2r}(1/2)^{[1/2 - 1/r]D/2 + D/4} = \left(e^{1/3}(1/2)^{2/3}\right)^{D/2}$$

when $r = 3$, the worst case. Since $4 > e$, the quantity in parentheses is < 1 when $\alpha = 0$ and hence also when $0 \le \alpha \le \alpha_0$, if α_0 is small.

To extend this result to other values of k, let

$$H(k) = \log G(k) = \frac{k}{r} + k\alpha \log \beta + k\gamma \log(k/D)$$
$$+ \frac{D - (1+\alpha)k}{2} \log \left(1 - \frac{(1+\alpha)k}{D}\right)$$

Since $G(k) = \exp(H(k))$, differentiating gives $G'(k) = G(k)H'(k)$ where

$$H'(k) = \frac{1}{r} + \alpha \log \beta + \gamma \log(k/D) + \gamma - \frac{(1+\alpha)}{2} \log \left(\frac{D - (1+\alpha)k}{D}\right)$$
$$+ \frac{D - (1+\alpha)k}{2} \cdot \frac{D}{D - (1+\alpha)k} \cdot \left(\frac{-(1+\alpha)}{2}\right)$$

Differentiating again $G''(k) = G(k)(H'(k)^2 + H''(k))$ where

$$H''(k) = \frac{\gamma}{k} - \frac{(1+\alpha)}{2} \cdot \frac{D}{D - (1+\alpha)k} \cdot \left(\frac{-(1+\alpha)}{D}\right) > 0$$

From the last calculation we see that $G(k)$ is convex. We have control of the value for $k = D/2$. It remains then to inspect the values for small k. Dropping the last factor which is < 1

$$G(k) \le e^{k/r} \beta^{k\alpha} \left(\frac{k}{D}\right)^{\gamma k}$$

When $0 \le \alpha \le \alpha_0 \le 1/24$, $\gamma \ge 7/48$ and hence $G(24) \le Cn^{-7/2}$. Since $e(S, S^c) \ge 1$ there is nothing to prove for $k \le 1/\alpha_0 = 24$ and the proof is complete. ∎

By using a simple comparison with the 3-regular graph, we will now prove a result for the random walk on the BC small world that stays put with probability $1/2$ and jumps to each of its three neighbors with probability $1/6$. We learned this from Elchanan Mossel during a meeting held at MSRI.

Theorem 6.3.4. *The random walk on the BC small world mixes in time $O(\log n)$.*

Proof. Suppose that our small world graph has $3n$ points. For reasons that will become clear in Case 2, we will estimate the conductance associated with $K^3(x, y)$. Let $A \subset \{1, 2, \ldots, 3n\}$ with $|A| \le 3n/2$, $I_j = \{3j - 2, 3j - 1, 3j\}$ for $1 \le j \le n$, $J = \{j : I_j \subset A\}$, $K = \{j \notin J : I_j \cap A \ne \emptyset\}$, and $B = \cup_{j \in J} I_j$.

Case 1. If $|J| \leq |A|/6$ then $|B| \leq |A|/2$ so $|K| \geq |A|/4$ and there are at least $|A|/4$ short-range edges connecting points in A to A^c. From this we conclude

$$\frac{Q(A, A^c)}{\pi(A)} \geq \frac{(1/24)(|A|/4)}{|A|}$$

Case 2. Let S be the small world graph and let $\mathcal{R}$ be the random 3-regular graph in which there is an edge from j to k for each edge from $x \in I_j$ to $y \in I_k$. This is a random multigraph which is the same as the random configuration model of Bollobás (1988). Since $|J| \leq n/2$, by a result in that paper there is a constant $\gamma > 0$ so that the number of edges connecting J to J^c satisfies $e(J, J^c) \geq \gamma|J|$. If there is an edge from j to k in $\mathcal{R}$ then there is an $x \in I_j$ that is at one end of the edge and a $y \in I_k$ with $y \notin A$. $p^3(x, y) > 1/216$ and $|J| \geq |A|/6$ so

$$\frac{Q(A, A^c)}{\pi(A)} \geq \frac{(1/216)e(J, J^c)}{|A|} \geq (1/216)\gamma/6$$

which completes the proof.   ∎

Both the random 3-regular graph and the BC small world look locally like a tree in which each vertex has degree 3. The lazy random walk on the tree moves further from the root with probability 2/6 and closer with probability 1/6, so the distance should increase like $t/6$. Berestycki and Durrett have shown that on the random 3-regular graph the distance of the random walk S_t from its starting point satisfies

$$d(0, S_{c \log_2 n}) \sim \left(\frac{c}{6} \wedge 1\right) \log_2 n$$

as $n \to \infty$ uniformly for c in compact sets. Since there are at most $n^{1-\epsilon}$ points within distance $(1 - \epsilon) \log_2 n$ of the origin, this shows that the mixing time is $\geq 6 \log_2 n$. Based on this one might

Conjecture 6.3.5. *The mixing time for the random walk on the random 3-regular graph is asymptotically* $6 \log_2 n$.

This has been proved by Lubetzky and Sly (2009).

6.4 Preferential Attachment Graph

Consider the Barabási–Albert preferential attachment graph $G_{n,d}$, defined in Section 4.1, in which each newly added vertex makes d connections to existing vertices. Introduce the lazy random walk on $G_{n,d}$ that stays put with probability $1/2$ and from x to each of its $d(x)$ neighbors with equal probability. The sum of the degrees of the graph is $2dn$ so the stationary distribution of this walk is $\pi(x) = d(x)/2dn$. Our proof follows Mihail, Papadimitriou, and Saberi (2004), but we need the following lemma to fix their proof.

Lemma 6.4.1. *There is a constant $a < 1$ so that if n is large and $\pi(S) \leq 1/2$ then with high probability $|S| \leq an$.*

Proof. To prove this we note that Theorem 4.1.4 implies that the fraction of vertices of degree k, $Z(k, n)/n \to p_k$. Pick K so that $\sum_{k=d}^{K} k p_k > d$ and set $a = \sum_{k=d}^{K} p_k$. To prove the claim now, note that to make the largest set with $\pi(S) \leq 1/2$ we should first take all of the vertices of degree $d, d + 1$, etc. until we have half of the volume. ∎

To bound the conductance it is useful to think of the edges as directed and pointing away from the vertex that was just added. From this we see that each vertex has out-degree d, so

$$\text{vol}(S) \leq 2d|S| + e(S, S^c)$$

since each edge incident to a vertex in S must be one of the $d|S|$ edges that come out from a vertex in S in which case it contributes at most 2 to the volume or it comes out from a vertex in S^c and is part of $e(S, S^c)$. Combining this with the lemma and the fact that $x/(2d + x)$ is increasing gives (recall $Q(S, S^c) = e(S, S^c)/2$ because the walk is lazy):

$$h \geq \frac{\iota_a}{2d + \iota_a}$$

where $\iota_a = \min_{S:|S| \leq an} e(S, S^c)/|S|$. We will find a positive lower bound for ι_a. It will then follow that

Theorem 6.4.2. *The mixing time of the lazy random walk on the Barabási–Albert preferential attachment with $d \geq 2$ graph is $O(\log n)$.*

To begin to estimate ι_a, recall that $G_{n,d}$ is constructed by first building the tree $G_{nd,1}$ and then identifying the mini-vertices $kd, kd - 1, \ldots (k - 1)d + 1$ to produce vertex k. Fix a set $S \subset \{1, 2, \ldots n\}$ and suppose $1 \notin S$. For this part of the argument we do not suppose $|S| \leq an$, so $1 \notin S$ can be achieved by relabeling S and S^c.

We start the construction with a single vertex 1 connected to itself, and count this self-loop as degree 1 for later preferential attachments. For $2 \leq t \leq dn$, define the father of t, $f(t)$ to be the vertex t' to which t connects. We set $f(1) = 1$. We say t is associated with S and write $t \to S$ if it is part of a vertex in S. We say t is good if either $t \to S$ and $f(t) \to S^c$ or $t \to S^c$ and $f(t) \to S$. Note that $1 = f(1)$ so 1 is always bad. The key to the proof is the following.

Lemma 6.4.3. *Suppose $|S| = k$.*

$$P(\text{the set of good mini-vertices is } A) \leq \binom{dk}{|A|} \bigg/ \binom{dn - |A|}{dk - |A|}$$

Proof. Let A_1 be the mini-vertices in A associated with S and A_2 be the mini-vertices associated with S^c. Let $k_1 = |A_1|$ and $k_2 = |A_2|$ and note that $k_1 + k_2 = |A|$.

Let $x_1 < x_2 < \ldots x_{dk-k_2}$ be the mini-vertices associated with S that do not belong to A. We may write $x_i = y_i + z_i + 1$ where y_i and z_i are the total number of mini-vertices that arrived prior to x_i and belong to A or A^c respectively.

Let $\bar{x}_1 < \bar{x}_2 < \ldots \bar{x}_{(dn-dk)-k_2}$ be the mini-vertices associated with S^c that do not belong to A. We may write $\bar{x}_i = \bar{y}_i + \bar{z}_i + 1$ where $\bar{y}_i$ and $\bar{z}_i$ are the total number of mini-vertices that arrived prior to $\bar{x}_i$ and belong to A or A^c respectively. It follows from these definitions that

$$\cup_{i=1}^{dk-k_1} \{z_i\} \cup \cup_{i=1}^{(dn-dk)-k_2} \{\bar{z}_i\} = \{0, 1, 2, \ldots dn - |A| - 1\} \tag{6.4.1}$$

The total volume of the graph when mini-vertex t arrives is $2(t-1) - 1$. If $t = x_i$ we can write this as $2(z_i + y_i) - 1$. When x_i arrives the total volume of S is due to (a) the $i - 1$ bad mini-vertices that arrived before x_i and are associated with S, each of which contribute degree 2, and (b) the y_i good mini-vertices that arrived before x_i and are associated with S, each of which contribute degree 1. Notice that $y_i \geq 1$ since $1 \notin S$ implies that the first mini-vertices in S belongs to A. Thus the total degree of S when x_i arrives is $2(i-1) + y_i$ and the probability x_i attaches to S (and hence is bad) is

$$\frac{2(i-1) + y_i}{2(z_i + y_i) - 1} \leq \frac{2(i-1) + y_i}{2z_i + y_i} \leq \frac{2i + 2|A|}{2z_i + 2 + 2|A|} = \frac{i + |A|}{z_i + 1 + |A|} \tag{6.4.2}$$

Here we have subtracted $y_i - 1 \geq 0$ from the denominator, added $-y_i + 2 + 2|A| \geq 0$ to numerator and denominator, and divided numerator and denominator by 2.

If $t = \bar{x}_i$ we can write the total volume of the graph as $2(t-1) - 1 = 2(\bar{z}_i + \bar{y}_i) - 1$. When $\bar{x}_i$ arrives the total volume of S^c is due to (a) the $i - 1$ bad mini-vertices that arrived before $\bar{x}_i$ and are associated with S^c, each of which contribute degree 2 except for mini-vertex 1 that contributes 1, and (b) the $\bar{y}_i$ good mini-vertices that arrived before x_i and are associated with S^c, each of which contribute degree 1. Thus the total degree of S^c when $\bar{x}_i$ arrives is $2(i-1) - 1 + \bar{y}_i$ and the probability $\bar{x}_i$ attaches to S^c (and hence is bad) is

$$\frac{2(i-1) - 1 + \bar{y}_i}{2(\bar{z}_i + \bar{y}_i) - 1} \leq \frac{2(i-1) - 1 + \bar{y}_i}{2\bar{z}_i + \bar{y}_i - 1} \leq \frac{2i + 2|A|}{2\bar{z}_i + 2 + 2|A|} = \frac{i + |A|}{\bar{z}_i + 1 + |A|} \tag{6.4.3}$$

Here we have subtracted $\bar{y}_i \geq 0$ from the denominator, added $-\bar{y}_i + 3 + 2|A| \geq 0$ to numerator and denominator, and divided numerator and denominator by 2.

Now (6.4.2) and (6.4.3) imply the probability that all of the mini-vertices in A^c are bad given that all of the mini-vertices in A are good is at most

$$\prod_{i=1}^{dk-k_1} \frac{i + |A|}{z_i + 1 + |A|} \prod_{i=1}^{(dn-dk)-k_2} \frac{i + |A|}{\bar{z}_i + 1 + |A|}$$

Using (6.4.1) the above is

$$= \prod_{i=1}^{dk-k_1} i + |A| \ \prod_{i=1}^{(dn-dk)-k_2} i + |A| \ \bigg/ \ \prod_{i=1}^{dn-|A|} i + |A|$$

$$= \frac{(dk - k_1 + |A|)!((dn - dk) - k_2 + |A|)!}{|A|!(dn)!} = \frac{(dk + k_2)!(dn - dk + k_1)!}{|A|!(dn)!}$$

where in the second step we have multiplied numerator and denominator by $(|A|!)^2$ and in the third we have used $k_1 + k_2 = |A|$. The above is

$$= \frac{(dk)!(dn - dk)!}{|A|!(dn - |A|)!} \cdot \prod_{i=0}^{k_2-1} \frac{dk + k_2 - i}{dn - i} \prod_{i=0}^{k_1-1} \frac{dn - dk + k_1 - i}{dn - k_2 - i}$$

The terms in the two products are ≤ 1. $dk + k_2 - i \leq dn - i$ follows from the fact that $d|S| + |A \cap S^c| \leq dn$, while $dn - dk + k_1 - i \leq dn - k_2 - i$ follows from $d|S^c| + |A \cap S| \leq dn$. Discarding the products the above is

$$\leq \frac{(dk)!(dn - dk)!}{|A|!(dn - |A|)!} = \frac{(dk - |A|)!(dn - dk)!}{(dn - |A|)!} \frac{(dk)!}{|A|!(dk - |A|)!}$$

$$= \binom{dk}{|A|} \bigg/ \binom{dn - |A|}{dk - |A|}$$

which completes the proof. ∎

With Lemma 6.4.3 in hand the rest is routine.

Lemma 6.4.4. *Suppose $d \geq 2$. There is a constant $\alpha > 0$ so that if n is large, $\iota_a \geq \alpha$ with high probability.*

Proof. There is nothing to prove for sets of size 1, so we suppose $|S| \geq 2$. The numerator on the right-hand side in Lemma 6.4.3 is an increasing function of $|A| \leq dk/2$, while the denominator $= \binom{dn-|A|}{dn-dk}$ is a decreasing function, so if $\alpha < d/2$,

$$P(\iota_a \leq \alpha) \leq \sum_{k=2}^{an} \binom{n}{k} \alpha k \binom{dn}{\alpha k} \frac{\binom{dk}{\alpha k}}{\binom{dn-\alpha k}{dk-\alpha k}}$$

The first factor gives the number of ways of picking S, the second takes account of the fact that there are αk possible values of $|A|$ to consider. The third term bounds the number of ways of picking A, while the final term is the bound from Lemma 6.4.3.

One way to pick $dk - \alpha k$ points from a set of size $dn - \alpha k$ is to pick k from a fixed subset of size n and the other $(d-1)k - \alpha k$ from the remaining $(d-1)n - \alpha k$ so

$$\binom{n}{k}\binom{(d-1)n - \alpha k}{(d-1)k - \alpha k} \le \binom{dn - \alpha k}{dk - \alpha k}$$

and it follows that our sum is

$$\le \sum_{k=2}^{an} \alpha k \binom{dn}{\alpha k}\binom{dk}{\alpha k}\binom{(d-1)n - \alpha k}{(d-1)k - \alpha k}^{-1}$$

Using $(n/k)^k \le \binom{n}{k} \le c(en/k)^k$ and the fact that if $b > a > c$ then $(a-c)/(b-c) \le a/b$ the above is

$$\le \sum_{k=2}^{an} \alpha k \left(\frac{edn}{\alpha k}\right)^{\alpha k} \left(\frac{edk}{\alpha k}\right)^{\alpha k} \left(\frac{(d-1)k - \alpha k}{(d-1)n - \alpha k}\right)^{(d-1)k - \alpha k}$$

$$\le \sum_{k=2}^{an} \alpha k \left(\frac{n}{k}\right)^{\alpha k} \left(\frac{ed}{\alpha}\right)^{2\alpha k} \left(\frac{k}{n}\right)^{(d-1)k - \alpha k}$$

$$= \sum_{k=2}^{an} G(k) \quad \text{where} \quad G(k) = \alpha k \left(\frac{ed}{\alpha}\right)^{2\alpha k} \left(\frac{k}{n}\right)^{(d-1)k - 2\alpha k}$$

Let $H(k) = \log G(k)$. $G'(k) = G(k)H'(k)$ where

$$H'(k) = \frac{1}{k} + 2\alpha \log\left(\frac{ed}{\alpha}\right) + (d - 1 - 2\alpha)\log(k/n) + (d - 1 - 2\alpha)$$

Differentiating again we have $G''(k) = G(k)(H'(k)^2 + H''(k))$ where

$$H''(k) = -1/k^2 + (d - 1 - 2\alpha)/k > 0$$

as long as $k \ge 2$ and $\alpha < (2d - 3)/4$. Here we need $d \ge 2$.

Since $G(k)$ is convex, $\max_{2 \le k \le an} G(k) \le \max\{G(2), G(an)\}$

$$H(an) = \log(\alpha an) + bn \quad \text{where} \quad b = 2\alpha a \log(ed/\alpha) + (d - 1 - 2\alpha)a \log(a)$$

The first term in b converges to 0 as $\alpha \to 0$ while the second one tends to $(d-1)$ $a \log(a) < 0$. Thus if we pick $\alpha > 0$ small we have $b < 0$. At the other end

$$G(2) = 2\alpha \left(\frac{ed}{\alpha}\right)^{4\alpha} \left(\frac{2}{n}\right)^{2(d - 1 - 2\alpha)}$$

$d \ge 2$ so if $\alpha < (d-1)/4$ the power in the last expression is > 1. Combining this with the other conclusion we have $\max_{2 \le k \le an} G(k) = o(n^{-1})$ and it follows that $P(\rho_a \le \alpha) \to 0$.   ∎

6.5 Connected Erdős–Rényi Graphs

In this section we will consider random walk on $ER(n, (c \log n)/n)$ with $c > 1$, which Theorem 2.8.1 has shown is connected with high probability for large n. We define the lazy random walk as in the previous sections. Let $d(x)$ be the degree of x, write $x \sim y$ if x and y are neighbors, and define a transition kernel by $K(x, x) = 1/2$, $K(x, y) = 1/2d(x)$ if $x \sim y$ and $K(x, y) = 0$ otherwise.

Theorem 6.5.1. *Consider $ER(n, (c \log n)/n)$ with $c > 1$. The lazy random walk mixes in time $O(\log n)$.*

Proof. Our proof follows Cooper and Frieze (2003b). We begin by estimating the maximum and minimum degrees of vertices.

Lemma 6.5.2. *There is a constant $\delta > 0$ so that if n is large then*

$$\delta c \log n \le d(x) \le 4c \log n \quad \text{for all } x$$

Proof. By the large deviations result Lemma 2.8.5 if $X = \text{Binomial}(n, p)$ then

$$P(X \ge np(1 + y)) \le \exp(-npy^2/2(1 + y))$$

Taking $p = (c \log n)/n$, and $y = 3$

$$P(X \ge 4c \log n) \le \exp(-9(c \log n)/8) = n^{-9c/8}$$

so with probability that tends to one, the maximum degree in the graph is $\le 4c \log n$.

To get a lower bound, we need the more precise result in Lemma 2.8.4. The function H defined there has $H(0) = -\log(1 - p)$, which is sensible since $P(X = 0) = (1 - p)^n$. When $p = c \log n/n$

$$(1 - (c \log n)/n)^n \le n^{-c}$$

Taking $a = (\delta c \log n)/n$, we have

$$H(a) = \frac{\delta c \log n}{n} \log \delta + \left(1 - \frac{\delta c \log n}{n}\right) \log \left(\frac{1 - \delta c \log n/n}{1 - c \log n/n}\right)$$

The logarithm in the second term is

$$\log \left(1 + \frac{(1 - \delta)c \log n/n}{1 - c \log n/n}\right) \sim (1 - \delta)c \log n/n$$

as $n \to \infty$. As $\delta \to 0$, $\delta \log \delta \to 0$, so if δ is small enough then $H(\delta p) \sim (b \log n)/n$ with $b > 1$ as $n \to \infty$ and we conclude that with probability that tends to one, the minimum degree in the graph is $\ge \delta c \log n$. ∎

To prove the theorem we will estimate the conductance h. By considering the number of edges we see that $\text{vol}(G)$ is $2\text{Binomial}(\binom{n}{2}, c\log n/n)$ which has mean $\sim cn\log n$ and variance $\sim cn\log n$, so $\text{vol}(G) \sim cn\log n$. The maximum degree $\leq 4c\log n$ with high probability for large n, and hence no set with $|S| \geq 9n/10$ will have $\pi(S) \leq 1/2$.

Case 1. Consider $B = \{S : n/(c\log n) \leq |S| \leq 9n/10\}$ and let $s = |S|$. Using Lemma 2.8.4 with $n = s(n-s)$, $p = p$, and $y = 1/2$

$$P(\exists S \in B : e(S, S^c) \leq s(n-s)p/2) \leq \binom{n}{s}\exp(-s(n-s)(c\log n)/8n)$$

Using Lemma 6.3.3 and $n - s \geq n/10$ the above is

$$\leq \exp\left(-s\left[\frac{c\log n}{80} - \log(n/s) - 1\right]\right)$$

$$\leq \exp\left(-\frac{n}{c\log n}\left[\frac{c\log n}{80} - \log(c\log n) - 1\right]\right)$$

which goes to 0 exponentially fast as $n \to \infty$. To finish up now we note that

$$s(n-s)p/2 = s(n-s)(c\log n)/2n \geq sc(\log n)/20$$

while $\text{vol}(S) \leq 4sc\log n$, so for sets in B we have $e(S, S^c)/\text{vol}(S) \geq 1/80$.

Case 2. $A = \{S : 1 \leq |S| \leq n/(c\log n)\}$. In this case we need an upper bound $e(S, S)$ in order to conclude $e(S, S^c)$ is large. $E|e(S, S)| \leq (s^2/2)p \leq s/2$ so

$$P(\exists S \in A : e(S, S) \geq s\log\log n) \leq C\binom{n}{s}\binom{s^2/2}{s\log\log n}p^{s\log\log n}$$

The right-hand side is the probability $e(S, S) = s\log\log n$, ignoring the fact that this may not be an integer. However in this part of the tail, the probabilities decay exponentially fast. Bounding the binomial coefficients using Lemma 6.3.3, and filling in the value of p

$$\leq C\left(\frac{ne}{s}\right)^s\frac{(s^2/2)^{s\log\log n}p^{s\log\log n}}{(s\log\log n)^{s\log\log n}e^{-s\log\log n}}$$

$$= C\left(\frac{ne}{s}\right)^s\left(\frac{s}{2\log\log n}\cdot\frac{ec\log n}{n}\right)^{s\log\log n}$$

Reorganizing we have

$$= C\exp\left(s[\log(ne) - \log s] + s\log\log n\left[\log s + \log\left(\frac{ec\log n}{2n\log\log n}\right)\right]\right)$$

Differentiating the exponent with respect to s we have

$$\log(ne) - \log s - 1 + \log\log n[\log s + \log(ec\log n) - \log(2n\log\log n)]$$
$$+ \log\log n$$

When $1 \le s \le n/(c\log n)$ this is negative, so the worst case is $s = 1$. In this case the quantity of interest is

$$= \exp\left(\log(ne) + \log\log n[\log(ec\log n) - \log(2n\log\log n)]\right)$$

which tends to 0 as $n \to \infty$.

To bound $e(S, S^c)$ we note that $e(S, S^c) = d(S) - e(S, S) \ge s\delta\log n - s\log\log n \ge s(\delta/2)\log n$ when n is large. $\mathrm{vol}(S) \le 4sc\log n$, so for sets in A we have $e(S, S^c)/\mathrm{vol}(S) \ge \delta/8$. ∎

6.6 Small Worlds

The BC small world was treated in Section 6.3. On the NW small world, the walker stays put with probability $1/2$ and with probability $1/2$ jumps to a random chosen neighbor. This leads to a stationary distribution $cd(x)$ where $d(x)$ is the degree of x. As in the previous section, to bound the mixing time we will estimate the conductance.

Given a set S with cardinality $|S| = s$ the number of shortcuts from S to S^c is Poisson with mean $ps(n-s)/n$. Thus if we have an interval of length $s = (1/p)\log n$ the probability of no shortcut is $\approx n^{-1}$, and we can expect to have one such interval in the small world. The time to escape from this interval starting from the middle is $O(\log^2 n)$ which gives a lower bound of $O(\log^2 n)$ on the mixing time.

When there is an interval of length $s = (1/p)\log n$, $h \le 2p/\log n$. Our goal is to prove that the conductance has

$$h \ge C/\log n \tag{6.6.1}$$

Combining this with our Markov chain results in Section 6.2 gives

Theorem 6.6.1. *Random walk on the NW small world mixes in time at least $O(\log^2 n)$ and at most $O(\log^3 n)$.*

Proof. The proof is similar to the one for connected Erdös–Renyi graphs but there are new problems arising from the fact that $\mathrm{vol}(S)$ may be as small as $2|S|$ or as large as $m|S|$ where $m = \log n/(\log\log n)$ is the maximum degree of vertices in the graph. To see the latter claim note that using Stirling's formula if $Z = \mathrm{Poisson}(p)$ then

$$P(Z = m) = e^{-p}p^m/m! = 1/n$$

when $m^m(ep)^{-m}e^p/\sqrt{2\pi m} = n$ or roughly when $m \log m = \log n$, that is, $m \sim \log n/(\log \log n)$.

Case 1 (Large sets). Suppose $|S| \geq n/\log n$. By Lemma 6.3.3 the number of sets of size $s = |S|$ is

$$\binom{n}{s} \leq \left(\frac{ne}{s}\right)^s = \exp\left(s\{\log(n/s) + 1\}\right)$$

To bound the volume of S we will generate independent Poisson mean p random variables ξ_x for each $x \in S$ and then connect them to points chosen independently and uniformly at random. This gives the right distribution for the number of long-range connections $\ell(S, S^c)$ but produces a Poisson mean $2p/n$ number of connections between $x, y \in S$. However, this is not a problem when we are interested in upper bounds on the volume vol(S). To be precise if $L(S) = \sum_{x \in S} \xi_x$ then vol(S) $\leq 2|S| + L(S) - \ell(S, S^c)$ in the sense of distribution.

By our large deviations result for the Poisson distribution, if $Z = \text{Poisson}(\mu)$ then

$$P(Z \geq y\mu) \leq e^{-\gamma(y)\mu} \quad \text{where} \quad \gamma(y) = y \log y - y + 1$$

When $y \geq e^2$, $\gamma(y) \geq (y \log y)/2$, so taking $\mu = 2ps$ we have

$$P(\text{vol}(S) \geq s \log \log n) \leq \exp(-ps \log \log n(\log \log \log n))$$

Since $1 + \log(n/s) \leq 2 \log \log n$ for $n \geq e^e$ it follows that if $s \geq n/\log n$ then the probability of a set of size s with vol(S) $\geq s \log \log n$ is

$$\leq \exp(-ps \log \log n(\log \log \log n - 2/p)) = o(n^{-1})$$

so with high probability we have vol(S) $\leq |S| \log \log n$ for all large sets.

Case 1a (Skinny large sets). Let $b > 0$. Orient the ring in some direction and define the left endpoints of a set to be the boundary points where the orientation points into the interval. If the number of left endpoints of S in the ring is $\geq s/(\log n)^b$ then

$$\frac{e(S, S^c)}{\text{vol}(S)} \geq \frac{2s/(\log n)^b}{s \log \log n} = \frac{2}{(\log n)^b \log \log n}$$

Case 1b (Fat large sets). Suppose now that the number of left endpoints, $r \leq s/(\log n)^b$. The number of sets of size s with this property is

$$\leq \binom{n}{r}\binom{s-1}{r-1}$$

where the second factor gives the number of ways of partitioning s into r pieces each of size ≥ 1. Note that in our clumsy counting we have done nothing to prevent the intervals from touching or overlapping. Thus all configurations with $r < s/(\log n)$ intervals can be written with $r = s/(\log n)$ intervals. Since

$$\binom{s-1}{r-1} = \binom{s-1}{s-r} \leq \binom{s}{s-r} = \binom{s}{r}$$

The number of sets is

$$\leq \left(\frac{ne}{r}\right)^r \cdot \left(\frac{se}{r}\right)^r = \exp\left(r\{\log(n/s) + 2\log(s/r) + 2\}\right) \tag{6.6.2}$$

When $r = s/(\log n)^b$ and $s \geq n/\log n$ this is

$$\leq \exp\left(\frac{s}{(\log n)^b}\{(1 + 2b)\log\log n + 2\}\right) \tag{6.6.3}$$

Note that if $\epsilon > 0$ and n is large the last quantity is $\leq \exp(\epsilon s)$, so it is enough to have large deviations estimates $\leq \exp(-\gamma s)$ for some $\gamma > 0$. To begin we note that a generalization of Lemma 6.4.1 implies that there is a constant $a > 0$ so that if n is large and $\pi(S) \leq 1/2$ then with high probability $|S| \leq (1-a)n$. By the construction above, the distribution of $\ell(S, S^c)$ conditional on $L(S) = L$ is larger than Binomial(L, a). Standard large deviations estimates for the Poisson and Binomial imply

$$P(\ell(S, S^c) < (a/2)L|L(S) = L) \leq \exp(-c(a)L)$$

$$P(L(S) \leq p|S|/2) \leq \exp(-c(p)|S|)$$

At this point there are two cases to consider. If $L(S) \geq |S|$ then vol$(S) \leq 4L(S)$ and

$$\frac{e(S, S^c)}{\text{vol}(S)} \geq \frac{\ell(S, S^c)}{4L(S)} \geq \frac{a}{8}$$

If $L(S) \leq |S|$ then vol$(S) \leq 4|S|$ and

$$\frac{e(S, S^c)}{\text{vol}(S)} \geq \frac{\ell(S, S^c)}{4|S|} \geq \frac{ap}{16}$$

Combining the last two results gives a positive lower bound on the conductance for fat large sets.

Case 2 (Small sets). By our construction, the distribution of $\ell(S, S)$ conditional on $L(S) = L$ is smaller than Binomial(L, p) where $p = s/n$. Using our Binomial large deviations result,

$$P(\ell(S, S) > L/2|L(S) = L) < \exp(-H(1/2)L)$$

Using $H(1/2) = (1/2)\log(1/2p) + (1/2)\log(1/2(1-p)) = (1/2)\log(1/4p(1-p)) > (1/2)\log(1/p) - 1$ and $1/p = n/s$.

$$P(\ell(S, S) \geq L/2 | L(S) = L) \leq \exp(-(L/2)(\log(n/s) - 2)) \qquad (6.6.4)$$

By Lemma 6.3.3, the number of sets of size $s = |S|$ is

$$\binom{n}{s} \leq \left(\frac{ne}{s}\right)^s = \exp(s\{\log(n/s) + 1\})$$

Combining the last two results, the probability of a set of size s with $L(S) \geq 4s$ and $\ell(S, S^c) \leq 2s$ is

$$\leq \exp(-s(\log(n/s) - 5))$$

When $s = 1$ this is $/1ne$ and for $s \geq 2$ this is $O(n^{-2})$, so for the rest of the proof we can suppose $L(S) \leq 4s$ and hence $\mathrm{vol}(S) \leq 6s$. Since $e(S, S^c) \geq 2$ when $\pi(S) \leq 1/2$, there is nothing to prove if $|S| < M\log n$, so for the rest of the proof we can suppose $|S| \geq M\log n$.

Case 2a (Skinny small sets). If $\mathrm{vol}(S) \leq 6s$ and the number of left endpoints is $\geq s/K(\log n)$ then

$$\frac{e(S, S^c)}{\mathrm{vol}(S)} \geq \frac{2s/K(\log n)}{6s} = \frac{1}{3K\log n}$$

Case 2b (Fat small sets). By (6.6.2) the number of sets is smaller with left endpoints $r \leq s/K(\log n)$ is

$$\leq \exp\left(\frac{s}{K\log n}\{\log(n/s) + 2\log(s/r) + 2\}\right)$$

Using (6.6.4) we have

$$P(\ell(S, S) \geq L/2 | L(S) = L) \leq \exp(-(L/2)(\log(n/s) - 2))$$

A standard large deviations estimate for the Poisson gives

$$P(L(S) \leq ps/2) \leq \exp(-c(p)s)$$

If $c(p)M > 1$ then the last quantity is $o(n^{-1})$. Combining the last two results gives a positive lower bound on the conductance for fat small sets, and completes the proof. ∎

Post mortem. For large sets we were able to show $h \geq 1/(\log n)^\epsilon$ for any $\epsilon > 0$. However, for small sets our bound was $C/\log n$. To see this cannot be improved, note that the probability of an interval of length $pr\log n$ with no long-range connections is n^{-r} so there will be roughly n^{1-r} intervals of this length and their union will be a set of size $prn^{1-r} \log n$ with $e(S, S^c) = 2n^{1-r}$.

6.7 Only Degrees 2 and 3

In this section we will consider a model that is closely related to an NW small world in which a fraction p of the sites in the ring have a long-range neighbor, and to the fixed degree distribution graphs with $p_3 = p$ and $p_2 = 1 - p$ but is easier to study. We start with a random 3-regular graph H with pn vertices, and produce a new graph G by replacing each edge by a path with a geometric number of edges with success probability r, that is, with probability $(1 - r)^{j-1}r$ we have j edges. The number of vertices of degree 2 in one of these paths has mean $(1/r) - 1$ so if we pick r so that $3p((1/r) - 1)/2 = 1 - p$, we asymptotically have the desired degree distribution. This is different from the usual fixed degree distribution since in that case the number of vertices of degree 2 is $(1 - p)n + O(1)$, while in the new model the number is $(1 - p)n + O(\sqrt{n})$.

Our main result is that

Theorem 6.7.1. *The mixing time of the lazy random walk on G is $O(\log^2 n)$.*

The lower bound follows as in the previous section from the fact that the longest path is $O(\log n)$. To prove an upper bound, we will use the Fountoulakis and Reed result. I learned this proof from Bruce Reed. Since we were drinking beer at the time, he should not be held responsible for my accounting of the details.

Lemma 6.7.2. *The number of connected subsets of H of size k containing a fixed vertex v_0 is $\le 3^{3k}$.*

Proof. Given a connected set V of vertices of H, define the set $W = \{(x, y) : x, y \in V, x \ne y\}$. Note that if $(x, y) \in W$ then $(y, x) \in W$ and think of these as two oriented edges between x and y. We will show that there is an Eulerian path starting from v_0 that traverses each oriented edge at least once. The number of edges in W is at most $3k$. At each stage we have at most three choices so the number of such paths is $\le 3^{3k}$ this proves the desired result.

To construct the path start at v_0 and pick an outgoing edge. When we are at a vertex $v \ne v_0$ we have used one more incoming edge than outgoing edge so we have at least one way out. This procedure may terminate by coming back to v_0 at a time when there are no more outgoing edges. If so, and we have not exhausted the graph, then there is some vertex v_1 on the current path with an outgoing edge. Repeat the construction starting from v_1 using edges not in the current path. We will eventually come back to v_1. We can combine the two paths by using the old path from v_0 to the first visit to v_1, using the new path to go from v_1 to v_1, and then the old path to return from v_1 to v_0. Repeating this construction we will eventually exhaust all of the edges. ∎

Let B be a connected subset of G, and let $A = B \cap H$. It is easy to see that A is a connected subset of H. By the isoperimetric inequality for random regular graphs, there is an $\alpha > 0$ so that $|\partial A| \geq \alpha |A|$, where ∂A is the set of edges (x, y) with $x \in A$ and $y \notin A$. From the construction of the graph it is easy to see that $|\partial B| = |\partial A|$.

It remains to see how big $|B|/|A|$ can be. When $|A| = 1$ we can have $|B| = O(\log n)$. The key to the proof is to show that the ratio cannot be big when $|A|$ is. Let X_i be i.i.d with $P(X_i = j) = (1 - r)^j r$ and let $S_m = X_1 + \cdots X_m$.

Lemma 6.7.3. *There are constants β and γ so that*

$$P(S_m \geq \beta \log n + \gamma m) \leq n^{-2}(2/81)^m$$

Proof. The moment generating function

$$\psi(\theta) = E e^{\theta X_i} = \sum_{j=0}^{\infty} (e^\theta (1 - r))^j r = \frac{r}{1 - e^\theta(1 - r)}$$

when $e^\theta(1 - r) < 1$. If we pick $\theta > 0$ so that $e^\theta(1 - r) = 1 - r/2$ then $\psi(\theta) = 2$. Markov's inequality implies

$$P(S_m \geq \beta \log n + \gamma m) \leq \psi(\theta)^m \exp(-\theta[\beta \log n + \gamma m])$$

Letting $\beta = 2/\theta$ and $\gamma = \log 81/\theta$ the desired result follows. ∎

If $|A| = k$ then the number of edges adjacent to some point in A is $\geq k + 2$, the value for a tree and $\leq 3k$. Since the number of connected sets of size k is $\leq n 27^k$ it follows that with probability $1 - O(n^{-1})$ we have $|B| \leq \beta \log n + 3\gamma |A|$ for connected sets B. From this it follows that

$$|\partial B| = |\partial A| \geq \alpha |A| \geq \frac{\alpha}{3\gamma}(|B| - \beta \log n)$$

if $|B| \geq 2\beta \log n$ then $|\partial B|/|B| \geq c$, while for $|B| \leq 2\beta \log n$, $|\partial B|/|B| \geq 2/|B|$. To evaluate $\int_{1/3n}^{1/2} dx/(x \Phi(x)^2)$ up to a constant factor we note that

$$\int_{2\beta \log n/n}^{1/2} \frac{dx}{x} = O(\log n)$$

while changing variables $y = nx, dy = n\,dx$ shows

$$\int_{1/3n}^{2\beta \log n/n} \frac{dx}{x(2/xn)^2} = \int_{1/3}^{2\beta \log n} (y/4)\,dy = O(\log^2 n)$$

and completes the proof.

Extension to Other Models. Consider the special case of the fixed degree distribution graph in which there are ℓ vertices of degree 2 and m of degree 3, where m is even. If we call this graph $G_{\ell,m}$ and look at the vertices of degree 3 in the graph and say that $i \sim j$ if they are connected by a path in which all other vertices have degree 2 then we get a 3-regular graph. H has $3m/2$ edges. Recalling that there are $b_{\ell,m} = \binom{\ell+3/2m-1}{3m/2-1}$ ways of partitioning ℓ objects into $3m/2$ groups of size ≥ 0, we see that each graph H can be generated by $b_{\ell,m}$ graphs $G_{\ell,m}$, so H is a random 3-regular graph with m vertices. Starting from H and using our procedure with geometric distributions, each $G_{\ell,m}$ has probability $(1 - r)^\ell r^{3m/2}$, so if we condition on L the total number of vertices inserted into edges, the result is uniform. If we have $m = np$ and pick r to make $EL = n(1 - p)$ then $P(L = n(1 - p)) = O(1/\sqrt{n})$. Since the error estimate in the proof above is $O(n^{-1})$, the result holds conditional on $L = \ell$. Having proved the result when there are ℓ vertices of degree 2 and m of degree 3, with $\ell/n \to (1 - p)$ and $m/n \to p$, it follows immediately that the result holds for an NSW graph in which the degrees d_i are i.i.d. with $P(d_i = 2) = 1 - p$ and $P(d_i = 3) = p$.

A similar argument applies to the NW_p small world. First, consider a ring in which an even number m of randomly chosen vertices have long-range neighbors. If we collapse the paths of vertices of degree 2 into edges then we have a BC small world on m vertices.

6.8 Hitting Times

We begin with the random walk on the BC small world. Pick two starting points x_1 and x_2 at random according to the stationary distribution π, which in this case is uniform, and define independent continuous time random walks X_t^1 and X_t^2 that jump at rate one and have $X_0^1 = x_1$ and $X_0^2 = x_2$. Let $A = \{(x, x)\}$ and $T_A = \inf\{t \geq 0 : X_t^1 = X_t^2\}$ be the first hitting time of A by (X_t^1, X_t^2). Writing π (rather than $\pi \times \pi$) for the initial condition $P_\pi(X_t^1 = X_t^2) = 1/n$, so we expect that T_A is $O(n)$. Indeed, if we notice that the expected amount of time $t \in [0, an]$ that $X_t^1 = X_t^2$ is a and if $X_t^1 = X_t^2$ for some $t \leq an - 1$ we will have $X_s^1 = X_s^2$ for $s \in [t, t + 1]$ with probability e^{-2} then

$$P_\pi(T_A \leq an - 1) \leq ae^2 \tag{6.8.1}$$

so if $a < e^{-2}$ then $P_\pi(T_A > an - 1)$ is bounded away from 0.

For the random walks on the NW small world or the fixed degree distribution graph, $P_\pi(X_t^1 = X_t^2) = \sum_x d(x)^2/D^2$ where $D = \sum_x d(x)$ is the sum of the degrees, so if $d(x)$ has finite second moment then as $n \to \infty$

$$\pi(A) \sim Ed(x)^2/n(Ed(x))^2$$

and again we can conclude that T_A takes a time at least $O(n)$. In the preferential attachment graph $Ed(x)^2 = \infty$ so we will not consider that graph here. Finally, in the connected Erdös–Rényi graphs the $d(x)$ are Poisson with mean $c \log n$, so $\sum_x d(x)^2/D^2 \sim 1/n$.

Proposition 23 of Aldous and Fill (2003) implies

$$\sup_t |P_\pi(T_A > t) - \exp(-t/E_\pi T_A)| \le \tau_2/E_\pi T_A \qquad (6.8.2)$$

where τ_2 is the relaxation time, which they define (see p. 19) to be 1 over the spectral gap. In all of our examples $\tau_2 \le C \log^2 n$ and as we will see

$$E_\pi T_A \sim cn \qquad (6.8.3)$$

so the hitting time is approximately exponential.

The proof of (6.8.2) is based on a result of Mark Brown (1983) for IMRL (increasing mean residual life) distributions. If one is willing to give up on the explicit error bound, it is fairly easy to give a proof based on the idea that since convergence to equilibrium occurs much faster than the two particles hitting, then subsequential limits of $T_A/E_\pi T_A$ must have the lack of memory property, and hence the sequence converges to a mean 1 exponential.

Theorem 6.8.1. *The mixing times of our chains $t_n = o(n)$, $n\pi(A) \to b$, and $E_\pi(T_A) \sim cn$, so under P_π, T_A/n converges weakly to an exponential with mean c.*

Proof. Let $\epsilon_n \to 0$ with $n\epsilon_n/t_n \to \infty$. By the argument for (6.8.1)

$$P_\pi(T_A \in [rn, (r + \epsilon_n)n]) \le e^2(\epsilon_n n + 1)\pi(A) \to 0$$

Using this result with $r = s + t$ and writing $X_t = (X_t^1, X_t^2)$, $x = (x_1, x_2)$

$$P_\pi(T_A > (s + t)n) = P_\pi(T_A > (s + t + \epsilon_n)n) + o(1)$$

$$= \sum_{x,y} P_\pi(T_A > sn, X_{sn} = x)P_x(X_{\epsilon_n n} = y)P_y(T_A > tn)$$

Subtracting $\pi(y)$ from $P_x(X_{\epsilon_n n} = y)$ and adding $\pi(y)$ gives two terms. The second is

$$\sum_{x,y} P_\pi(T_A > sn, X_{sn} = x)\pi(y)P_y(T_A > tn) = P_\pi(T_A > sn)P_\pi(T_A > tn)$$

The absolute value of the first term is bounded by

$$P_\pi(T_A > sn) \sup_x \sum_y |P_x(X_{\epsilon_n n} = y) - \pi(y)| \le \Delta(\epsilon_n n) \to 0$$

Since $E_\pi T_A \sim cn$, the sequence T_A/n is tight. Let F denote a subsequential limit. From the calculation above we see that if s, t, and $s + t$ are continuity points then

$$1 - F(s + t) = (1 - F(s))(1 - F(t))$$

F can have at most countably many discontinuity points, so there is a $\theta > 0$ so that F is continuous at all points $m/(\theta 2^n)$ where m and n are positive integers. Define λ by $e^{-\lambda} = 1 - F(1/\theta)$. It follows from the equation that if $t = m/(\theta 2^n)$ then $1 - F(t) = e^{-\lambda t}$. To conclude that λ is independent of the subsequential limit, note that for large n, $P(T_A > 3cn) \le 1/2$, so using the calculation above we see that if $\epsilon > 0$ and n is large

$$P(T_A > (k + 1)(3cn)) \le P(T_A > k(3cn))(\epsilon + P(T_A > 3cn))$$

This gives an exponential bound on the tail of the distribution, which enables us to conclude that every subsequential limit has mean c, that is, the constant λ is always $1/c$ and the proof is complete. ∎

Proof of (6.8.3). We use a version of Aldous' Poisson clumping heuristic. Consider the discrete time version $\bar{X}_n$ of the two particle chain in which at each step we pick a particle at random and let it jump. Writing P_A for $P_\pi(\cdot|X_0 \in A)$, a theorem of Kac implies that

$$E_A(T_A) = 1/\pi(A)$$

Starting from the diagonal, the two particles may hit in a time that is $O(t_n)$. The expected value on this event makes a contribution that is $o(n)$ to the expected value. When the two particles don't hit in $O(t_n)$ the chain is close to equilibrium, so

$$1/\pi(A) \approx P_A(T_A \gg t_n)E_\pi(T_A)$$

and we have

$$E_\pi(T_A) = \frac{1}{\pi(A)} \cdot \frac{1}{P_A(T_A \gg t_n)}$$

To connect with the clumping heuristic, we note that the naive guess for the waiting time is $1/\pi(A)$ but this must be corrected for by multiplying by the clump size, that is, the expected number of hits that occur soon after the first one. In nice cases, for example, the BC small world, the number of hits is a geometric with mean $1/P_A(T_A \gg t_n)$.

The value of $P_A(T_A \gg t_n)$ depends on the example but in each case, it comes from thinking about what the graph looks like locally.

BC small world. Locally the BC small world looks like a tree in which each vertex has degree 3. The probability that two random walkers that start from the origin will hit after they separate is the same as the probability that a single random walk will return to the origin, which is $1/2$.

NW small world. The first step is to describe what the space looks like locally. To do this we use the two type branching process from Section 5.3 with red and blue particles that correspond to short- and long distance neighbors. The process starts with one blue particle. All particles have a Poisson mean p number of blue offspring. Blue particles always have two red offspring, while red particles always have one. Having built the graph, we start both particles and let $e(\omega)$ be the probability that the two particles never hit after they separate. Recalling that $P_\pi(X_0^1 = X_0^2) = \sum_x d(x)^2/D^2$, we see that the desired constant is the expected value of $e(\omega)$ under biased measure in which the root has degree k with probability proportional to $k^2 P(Z = k)$ where Z is 2 plus a Poisson mean p.

The *fixed degree distribution* is similar to the NW small world and is left as an exercise for the reader.

In the *Connected Erdös–Rényi graphs* the degrees tend to infinity so $P_A(T_A \gg t_n) \to 1$.   ∎

In our four examples $E_\pi T_A \sim cn$. Kac's result implies that $E_A T_A = 1/\pi(A)$ and we have computed that

$$\pi(A) = \sum_{x \text{ in } A} d(x)^2/D^2 \quad \text{where} \quad D = \sum_x d(x)$$

When the degree distribution has infinite variance $\pi(A)$ will go to 0 more slowly than $1/n$. Consider random graphs with a fixed degree distribution $p_k \sim Ck^{-\nu}$. When $\nu > 3$ the distribution has finite variance. For $2 \le \nu \le 3$ results of Sood and Redner (2005) for the consensus time of the voter model, which we will consider in the next section, suggest that

$$E_\pi T_A \sim \begin{cases} n/\log n & \nu = 3 \\ n^{(2\nu-4)/(\nu-1)} & 2 < \nu < 3 \\ (\log n)^2 & \nu = 2 \end{cases}$$

We leave it to the reader to check that the right-hand side gives the behavior of $1/\pi(A)$.

6.9 Voter Models

The voter model was introduced independently by Clifford and Sudbury (1973) and Holley and Liggett (1975) on the d-dimensional integer lattice. It is a very simple model for the spread of an opinion and has been investigated in great detail, see Liggett (1999) for a survey. On any of our random graphs it can be defined as follows. Each site x has an opinion $\xi_t(x)$ and at the times T_n^x, $n \geq 1$ of a rate 1 Poisson process decides to change its opinion. To do this it picks a neighbor $y_{n,x}$ at random, and at time $t = T_n^x$ we set $\xi_t(x) = \xi_t(y_{n,x})$.

To analyze this process we use a "dual process" that works backwards in time to determine the source of the opinion at x at time t. To define this process we place a dot at x at time T_n^x and draw an arrow from (x, T_n^x) to $(y_{n,x}, T_n^x)$. To define $\zeta_s^{x,t}$ we start with $\zeta_0^{x,t} = x$. The process stays at x until the first time s that there is a dot at x. If this occurs at time $t - s = T_n^x$ then $\zeta_s^{x,t} = y_{n,x}$ and we continue to work our way down until we encounter a dot. This definition guarantees that the opinion of x at time t is the same as that of $\zeta_s^{x,t}$ at time $t - s$.

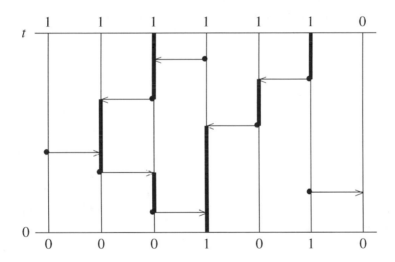

For fixed x and t, $\zeta_s^{x,t}$ is a random walk that jumps at rate 1 and to a neighbor chosen at random. It should be clear from the definition that if $\zeta_s^{x,t} = \zeta_s^{y,t}$ for some s then the two random walks will stay together at later times. For these reason the $\zeta_s^{x,t}$ are called coalescing random walks. If we consider the voter model on $\mathbb{Z}^d$ with the usual nearest neighbors then as Holley and Liggett (1975) have shown the recurrence of random walks in $d \leq 2$ and transience in $d > 3$ implies

Theorem 6.9.1. *In $d \leq 2$ the voter model approaches complete consensus, that is, $P(\xi_t(x) = \xi_t(y)) \to 1$. In $d > 3$ if we start from product measure with density p*

(i.e., we assign opinions $\bar{1}$ and $\bar{0}$ independently to sites with probabilities p and $1 - p$) then as $t \to \infty$, ξ_t^p converges in distribution to ξ_∞^p, a stationary distribution in which a fraction p of the sites have opinion 1.

On a finite set the voter model will eventually reach an absorbing state in which all voters have the same opinion. Cox (1989) studied the voter model on a finite torus $(\mathbb{Z} \bmod N)^d$ and showed:

Theorem 6.9.2. *Let ξ_t^p denote the voter model starting from product measure with density $p \in (0, 1)$. The time to reach consensus τ_N satisfies*

$$\tau_N = O(s_N) \quad \text{where} \quad s_N = \begin{cases} N^2 & d = 1 \\ N^2 \log N & d = 2 \\ N^d & d \geq 3 \end{cases}$$

and $E\tau_N \sim c_d[-p \log p - (1 - p) \log(1 - p)]s_N$, where c_d is a constant that depends on the dimension. In $d \geq 3$ the finite system looks like the stationary distribution for the infinite system at times that are large but $o(s_N)$.

As the next result shows, the voter model on many of our random graphs has the $d \geq 3$ behavior.

Theorem 6.9.3. *In the voter model on any of the random graphs considered in the previous section, the consensus time for ξ_t^p, $p \in (0, 1)$ is asymptotically at least $c_p n$ where $c_p > 0$ and n is the number of vertices in the graph.*

Proof. By Theorem 6.8.1 the hitting time of two randomly chosen points is $O(n)$. When these two random walks have not coalesced, their starting sites will be different with probability $2p(1 - p)$. ∎

It is natural to conjecture that the consensus time will be asymptotically $c_G n$ where c_G is a constant that depends on the random graph. However, in order to prove this we would have to understand the behavior of the coalescing random walk starting from all sites occupied. The next result is a first step in that direction. For simplicity, we consider only the easiest model.

Theorem 6.9.4. *Consider the BC small world and put m walkers on the graph at random starting points. The number of particles in the coalescing random walk at time nt converges to Kingman's coalescent in which transitions from k to $k - 1$ occur at rate $\binom{k}{2}$.*

Proof. The proof requires a number of messy details, so we will be content to describe the ideas. Let W_t^i be independent random walks. Let

$$A_{i,j,k,\ell} = \{|W_t^i - W_t^j| \text{ and } |W_t^k - W_t^\ell| \le (1/10)\log_2 n\}$$
$$B_{i,j,k} = \{|W_t^i - W_t^j| \text{ and } |W_t^i - W_t^k| \le (1/10)\log_2 n\}$$

where all the indices are distinct. Using π for the initial state in which m particles are scattered at random on the graph,

$$P_\pi(A_{i,j,k,\ell}) \le n \cdot n^{1/10} \cdot n \cdot n^{1/10}/n^4 = n^{-1.8}$$
$$P_\pi(B_{i,j,k}) \le n \cdot n^{1/10} \cdot n^{1/10}/n^3 = n^{-1.8}$$

so with high probability at all times $\le n \log n$ there are never two pairs that are closer than $(1/10)\log_2 n$.

Suppose that at some point, two particles are separated by $(1/10)\log_2 n$. We know that with high probability the graph seen in a neighborhood of radius $(1/3)\log_2 n$ around any point has at most one edge that deviates from the tree with degree 3. For simplicity we will ignore this annoying detail and suppose that the graph looks exactly like the tree with degree 3. In this case the distance d between the two points increases by 1 with probability $2/3$ and decreases by 1 with probability $1/3$. 2^{-d} is a harmonic function for this random walk, so a simple application of the optional stopping theorem shows that with probability $\le n^{-1/10}$ the distance will hit 0 before it increases to $(3/10)\log_2 n$. At this point it is likely that each particle will have moved about $(1/10)\log_2 n$ from its starting point so both are still in the original viewing window, and our tree assumption is still valid.

The last argument shows that if two particles are at distance $(1/10)\log_2 n$ then with probability $\ge 1 - n^{-2/10}$ they will separate to distance $(3/10)\log_2 n$ before they hit. The separation takes time $\ge (1/10)\log_2 n$ with high probability (the expected time is $(3/20)\log_2 n$ since each particle is moving away at rate $1/3$). This shows that with high probability the hitting of these two particles will take time at least $n^{1/10}$, at which time the positions of the particles have again randomized.

The last argument shows that the successive hitting times are roughly independent. The final step is to argue that when there are k particles, the hitting of the various pairs are roughly independent so the coalescence rate is $\binom{k}{2}$. The calculation follows the approach introduced by Cox and Griffeath (1986) and used by Cox (1989). Let τ be the time of the first coalescence, and τ_{ij} be the hitting time of particles i and j. Let $H_t(i, j) = \{\tau_{ij} \le nt\}$, $F_t(i, j) = \{\tau = \tau_{ij} \le nt\}$, and $q(t) = P(\tau \le nt)$.

$$P(H_t(i, j)) = P(F_t(i, j)) + \sum_{\{k,\ell\} \ne \{i,j\}} \int_0^{nt} P(\tau = \tau_{k\ell} = s, \tau_{ij} \le nt)\, ds$$

where the quantity being integrated on the right is the density of the hitting time. To evaluate the k, ℓ term in the sum we break things down according to the locations W_s^i and W_s^j. By our first observation we can ignore the possibility that $|W_s^i - W_s^j| < (1/10) \log_2 n$. When the distance is $\geq (1/10) \log_2 n$ our argument shows that the positions will become randomized before they hit so the hitting time will be in the limit as $n \to \infty$ exponentially distributed with mean n. Writing ϵ_n for a quantity that goes to 0 as $n \to \infty$

$$\int_0^{nt} P(\tau = \tau_{k\ell} = s, \tau_{ij} \leq nt)\, ds = \int_0^{nt} P(\tau = \tau_{k\ell} = s)(1 - \exp(-t + (s/n)))\, ds + \epsilon_n$$

Integrating by parts and then changing variables, the above is

$$= \int_0^{nt} \frac{1}{n} \exp(-t + (s/n)) P(\tau = \tau_{k\ell} \leq s)\, ds + \epsilon_n$$

$$= \int_0^{t} \exp(-(t - u)) P(\tau = \tau_{k\ell} \leq un)\, du + \epsilon_n$$

Using this in our initial decomposition, with the convergence of the hitting time of i and j to the exponential distribution

$$1 - e^{-t} = P(F_t(i, j)) + \sum_{\{k,\ell\} \neq \{i,j\}} e^{-t} \int_0^{t} e^{s} P(F_s(k, \ell))\, ds + \epsilon_n \qquad (6.9.1)$$

Summing over all $\binom{k}{2}$ pairs

$$\binom{k}{2}(1 - e^{-t}) = q(t) + \left[\binom{k}{2} - 1\right] e^{-t} \int_0^{t} e^{s} q(s)\, ds + \epsilon_n$$

It follows [see page 365 of Cox and Griffeath (1986) for details] that as $n \to \infty$, $q(t)$ converges to $u(t)$ the solution of

$$\binom{k}{2}(1 - e^{-t}) = u(t) + \left[\binom{k}{2} - 1\right] e^{-t} \int_0^{t} e^{s} u(s)\, ds$$

Multiplying both sides by e^{t} and rearranging we have

$$e^{t} u(t) - \binom{k}{2}(e^{t} - 1) = -\left[\binom{k}{2} - 1\right] \int_0^{t} e^{s} u(s)\, ds$$

Differentiating we have

$$e^{t} u(t) + e^{t} u'(t) - \binom{k}{2} e^{t} = -\left[\binom{k}{2} - 1\right] e^{t} u(t)$$

Dividing by e^t and rearranging gives

$$\binom{k}{2}(1 - u(t)) = u'(t) = -\frac{d}{dt}(1 - u(t))$$

which has solution $1 - u(t) = \exp(-t\binom{k}{2})$.

The final detail is to show that all $\binom{k}{2}$ pairs have equal probability to be the next to coalesce. To do this we go back to (6.9.1) and add and subtract $P(F_s(i, j))$ inside the integral to get

$$1 - e^{-t} = P(F_t(i, j)) - e^{-t}\int_0^t e^s P(F_s(i, j))\, ds + e^{-t}\int_0^t e^s q(s)\, ds + \epsilon_n$$

It follows that $P(F_t(i, j))$ converges to the solution of

$$v(t) - e^{-t}\int_0^t e^s v(s)\, ds = 1 - e^{-t} - e^{-t}\int_0^t e^s u(s)\, ds$$

Since the limit is independent of i, j we must have $v(t) = u(t)/\binom{k}{2}$, which completes the proof. ∎

A Special Voter Model. Consider the voter model defined by picking an edge at random from the graph, flipping a coin to decide on an orientation (x, y), and then telling the voter at y to imitate the voter at x. This version of the voter model has a uniform stationary distribution and in the words of Suchecki, Eguíluz, and Miguel (2005): "conservation of the global magnetization." In terms more familiar to probabilists, the number of voters with a given opinion is a time change of simple random walk and hence is a martingale. If we consider the biased voter model in which changes from 0 to 1 are always accepted but changes from 1 to 0 occur with probability $\lambda < 1$, then the last argument shows that the number of voters with a given opinion is a time change of a biased simple random walk and hence the fixation probability for a single 1 introduced in a sea of 0's does not depend on the structure of the graph. This is the small world version of a result of Maruyama (1970) and Slatkin (1981), which has been generalized by Lieberman, Hauert, and Nowak (2005).

To use our Markov chain results, we need a discrete time chain. Let $M = \max d(x)$ and define $p(x, x) = 1 - d(x)/M$ and $p(x, y) = 1/M$ when y is a neighbor of x. This is the embedded chain – at times of a rate M Poisson process, our walk takes a jump according to p. We will follow Lieberman, Hauert, and Nowak (2005) and call this the *isothermal random walk*. Since for each pair of neighbors $p(x, y) = 1/M$, the stationary distribution is uniform and the conductance is

$$h = \min_{S: |S| \le n/2} \frac{e(S, S^c)/M}{|S|}$$

It follows from (6.6.1) that

Theorem 6.9.5. *The isothermal random walk on the NW small world mixes in time at least* $O(\log^2 n)$ *and at most* $O(\log^4 n / \log \log n)$.

Proof. The conductance is $\geq C/(M \log n)$ so the spectral gap is $\geq C/(M \log n)^2$ and the convergence time for the discrete chain is $\leq C/(M^2 \log^3 n)$. The continuous time chain runs at M times the rate of the discrete time chain so its convergence time is $\leq C/(M \log^3 n)$. As we observed at the beginning of Section 6.6, $M \sim \log n/(\log \log n)$. ■

Remark. Suchecki, Eguuíluz, and Miguel (2005) have investigated the voter model on the Barabási–Albert scale-free network. They report that the time to consensus for the usual voter model, with node updates, scales as $n^{0.88}$, while for the isothermal voter model with edge updates the consensus time scales as n.

7

CHKNS Model

7.1 Heuristic Arguments

Inspired by Barabási and Albert (1999), Callaway, Hopcroft, Kleinberg, Newman, and Strogatz (2001) introduced the following simple version of a randomly grown graph. Start with $G_1 = \{1\}$ with no edges. At each time $n \geq 2$, we add one vertex and with probability δ add one edge between two randomly chosen vertices. Note that the newly added vertex is not necessarily an endpoint of the added edge and when n is large, it is likely not to be.

The CHKNS analysis of their model begins by examining $N_k(t) =$ the expected number of components of size k at time t. Ignoring terms of $O(1/t^2)$, which come from picking the same cluster twice:

$$N_1(t+1) = N_1(t) + 1 - 2\delta \frac{N_1(t)}{t}$$

$$N_k(t+1) = N_k(t) - 2\delta \frac{k N_k(t)}{t} + \delta \sum_{j=1}^{k-1} \frac{j N_j(t)}{t} \cdot \frac{(k-j) N_{k-j}(t)}{t}$$

To explain the first equation and for $k \geq 2$, note that at each discrete time t one new vertex is added, and a given isolated vertex becomes the endpoint of an added edge with probability $\approx 2\delta/t$. For the second equation, note that the probability an edge connects to a given cluster of size k is $\approx 2\delta k/t$, while the second term corresponds to mergers of clusters of size j and $k - j$. There is no factor of 2 in the last term since we sum from 1 to $k - 1$.

Theorem 7.1.1. *As $t \to \infty$, $N_k(t)/t \to a_k$ where $a_1 = 1/(1 + 2\delta)$ and for $k \geq 2$*

$$a_k = \frac{\delta}{1 + 2\delta k} \sum_{j=1}^{k-1} j a_j \cdot (k - j) a_{k-j}$$

Proof. The first equation has the form in Lemma 4.1.1 with $c = 1$ and $b = 2\delta$. The kth equation has the form in Lemma 4.1.2 with $b = 2\delta k$ and

$$g(t) = \delta \sum_{j=1}^{k-1} \frac{jN_j(t)}{t} \cdot \frac{(k-j)N_{k-j}(t)}{t}$$

which has a limit by induction. ∎

To solve for the a_k, which gives the limiting number of clusters of size k per site, CHKNS used generating functions. Let $h(x) = \sum_{k=1}^{\infty} x^k a_k$ and $g(x) = \sum_{k=1}^{\infty} x^k k a_k$. Multiplying the equations in Theorem 7.1.1 by $(1 + 2\delta k)x^k$, recalling $k = 1$ is different from the others, and summing gives

$$h(x) + 2\delta g(x) = x + \delta g^2(x)$$

Since $h'(x) = g(x)/x$ differentiation gives $g(x)/x + 2\delta g'(x) = 1 + 2\delta g(x)g'(x)$. Rearranging we have $2\delta g'(x)(1 - g(x)) = 1 - g(x)/x$ and

$$g'(x) = \frac{1}{2\delta x} \cdot \frac{x - g(x)}{1 - g(x)} \qquad (\star)$$

Let $b_k = ka_k$ be the fraction of vertices that belong to clusters of size k; $g(1) = \sum_{k=1}^{\infty} b_k$ gives the fraction of vertices that belong to finite components; $1 - g(1)$ gives the fraction of sites that belong to clusters whose size grows in time. Even though it is not known that the missing mass in the limit belongs to a single cluster, it is common to call $1 - g(1)$ the fraction of sites that belong to the giant component. The next result gives the mean size of finite components.

Lemma 7.1.2.

(i) If $g(1) < 1$ then $\sum_{k=1}^{\infty} kb_k = g'(1) = 1/2\delta$.

(ii) If $g(1) = 1$ then $g'(1) = (1 - \sqrt{1 - 8\delta})/4\delta$.

Proof. The first conclusion is immediate from $(\star)$. If $g(1) = 1$, L'Hôpital's rule implies

$$2\delta g'(1) = \lim_{x \to 1} \frac{x - g(x)}{1 - g(x)} = \lim_{x \to 1} \frac{1 - g'(x)}{-g'(x)}$$

which gives $2\delta(g'(1))^2 - g'(1) + 1 = 0$. The solution of this quadratic equation indicated in (ii) is the one that tends to 1 as $\delta \to 0$. ∎

Claim 7.1.3. *The critical value* $\delta_c = \sup\{\delta : g(1) = 1\} = 1/8$ *and hence*

$$\sum_k kb_k = \begin{cases} (1 - \sqrt{1 - 8\delta})/4\delta & \delta \leq 1/8 \\ 1/2\delta & \delta > 1/8 \end{cases}$$

Note that this implies that the mean cluster size $g'(1)$ is always finite, but is discontinuous at $\delta = 1/8$, since the value there is 2 but the limit for $\delta \downarrow 1/8$ is 4.

Physicist's proof. The formula for the derivative of the real-valued function g becomes complex for $\delta > 1/8$, so we must have $\delta_c \leq 1/8$, or Lemma 7.1.2 would give a contradiction. This conclusion is rigorous, but to argue the other direction, CHKNS note that mean cluster size $g'(1)$ is in general nonanalytic only at the critical value, and $(1 - \sqrt{1 - 8\delta})/4\delta$ is analytic for $\delta < 1/8$. If you are curious about their exact words, see the paragraph above (17) in their paper. ∎

To investigate the size of the infinite component, CHKNS integrated the differential equation ($\star$) near $\delta = 1/8$. Letting $S(\delta) = 1 - g(1)$ the fraction of vertices in the infinite component they plotted $\log(-\log S)$ versus $\log(\delta - 1/8)$ and concluded that

$$S(\delta) \sim \exp(-\alpha(\delta - 1/8)^{-\beta})$$

where $\alpha = 1.132 \pm 0.008$ and $\beta = 0.499 \pm 0.001$. Based on this they conjectured that $\beta = 1/2$.

Inspired by their conjecture Dorogovstev, Mendes, and Samukhin (2001) calculated that as $\delta \downarrow 1/8$,

$$S \equiv 1 - g(1) \approx c \exp(-\pi/\sqrt{8\delta - 1}) \tag{7.1.1}$$

This result shows $\beta = 1/2$ and $\alpha = \pi/\sqrt{8} = 1.1107$. Note that this implies that the percolation probability S is infinitely differentiable at the critical value, in contrast to the situation for the Erdős–Rényi model and for percolation on the small world in which $S \sim (\delta - \delta_c)$ as $\delta \downarrow \delta_c$.

Semi-rigorous proof. To derive this result Dorogovtsev, Mendes, and Samukhin (2001) change variables $u(\xi) = 1 - g(1 - \xi)$ in ($\star$) to get

$$u'(\xi) = \frac{1}{2\delta(1 - \xi)} \cdot \frac{u(\xi) - \xi}{u(\xi)}$$

They discard the $1 - \xi$ in the denominator (without any justification or apparent guilt at doing so) and note that the solution to the differential equation is the solution of the following transcendental equation

$$-\frac{1}{\sqrt{8\delta - 1}}\arctan\left(\frac{4\delta[u(\xi)/\xi] - 1}{\sqrt{8\delta - 1}}\right) - \ln\sqrt{\xi^2 - u(\xi)\xi + 2\delta u^2(\xi)}$$

$$= -\frac{\pi/2}{\sqrt{8\delta - 1}} - \ln\sqrt{2\delta} - \ln S$$

This formula is not easy (for me at least) to guess, although others more skilled with differential equations tell me it is routine. In any case with patience it is not hard to verify. Once this is done, the remainder of the proof is fairly routine asymptotic analysis of the behavior of the formula above as $\xi \to 1$. This could be cleaned up with some effort. The real mystery is why can one drop the $1 - \xi$?   ∎

7.2 Proof of the Phase Transition

In the original CHKNS model the number of edges added at each step is 1 with probability δ, and 0 otherwise. To obtain a model that we can analyze rigorously, we will study the situation in which a Poisson mean δ number of edges are added at each step. We prefer this version since in the Poisson case if we let $A_{i,j,k}$ be the event no (i, j) edge is added at time k then $P(A_{i,j,k}) = \exp\left(-\delta/\binom{k}{2}\right)$ for $i < j \le k$ and these events are independent.

$$P\left(\cap_{k=j}^n A_{i,j,k}\right) = \prod_{k=j}^n \exp\left(-\frac{2\delta}{k(k-1)}\right)$$

$$= \exp\left(-2\delta\left(\frac{1}{j-1} - \frac{1}{n}\right)\right) \ge 1 - 2\delta\left(\frac{1}{j-1} - \frac{1}{n}\right) \qquad \#1$$

The last formula is not simple, so we will also consider two approximations

$$\approx 1 - 2\delta\left(\frac{1}{j} - \frac{1}{n}\right) \qquad\qquad\qquad\qquad \#2$$

$$\approx 1 - \frac{2\delta}{j} \qquad\qquad\qquad\qquad\qquad\qquad \#3$$

We will refer to these three models by their numbers, and the original CHKNS model as #0. The approximation that leads to #3 is not as innocent as it looks. If we let $\mathcal{E}_n$ be the number of edges at time n then using the definition of the model $E\mathcal{E}_n \sim \delta n$ in models #0, #1, and #2 but $E\mathcal{E}_n \sim 2\delta n$ in model #3. It turns out, however, that despite having twice as many edges, the connectivity properties of model #3 is almost the same as that of models #1 and #2.

Theorem 7.2.1. *In models #1, #2, or #3, the critical value $\delta_c = 1/8$.*

In contrast to the situation with ordinary percolation on the square lattice where Kesten (1980) proved the physicists' answer was correct nearly 20 years after they had guessed it, this time the rigorous answer predates the question by more than 10 years. We begin by describing earlier work on the random graph model on $\{1, 2, 3, \ldots\}$ with $p_{i,j} = \lambda/(i \vee j)$. Kalikow and Weiss (1988) showed that the probability G is connected (ALL vertices in ONE component) is either 0 or 1, and that $1/4 \le \lambda_c \le 1$. They conjectured $\lambda_c = 1$ but Shepp (1989) proved $\lambda_c = 1/4$. To connect with the answer in Theorem 7.2.1, note that $\lambda = 2\delta$. Durrett and Kesten (1990) proved a result for a general class of $p_{i,j} = h(i, j)$ that are homogeneous of degree -1, that is, $h(ci, cj) = c^{-1}h(i, j)$. It is their methods that we will use to prove the result.

Proof of $\delta_c \ge 1/8$. We prove the upper bound for the largest model, #3. An easy comparison shows that the mean size of the cluster containing a given point i is bounded above by the expected value of the total progeny of a discrete time multitype branching process in which a particle of type j gives birth to one offspring of type k with probability $p_{j,k}$ (with $p_{j,j} = 0$) and the different types of births are independent.

To explain why we expect this comparison to be accurate, we note that in the Erdös–Rényi random graph with $p_{j,k} = \lambda/n$, the upper bound is an ordinary branching process with a Poisson mean λ offspring distribution, so we get the correct lower bound $\lambda_c \ge 1$. When $p_{j,k} = 2\delta/(j \vee k)$, the mean of the total progeny starting from one individual of type i is $\sum_{m=0}^{\infty} \sum_j p_{i,j}^m$, which will be finite if and only if the spectral radius $\rho(p_{i,j}) < 1$. By the Perron–Frobenius theory of positive matrices, ρ is an eigenvalue with positive eigenvector.

Following Shepp (1989) we now make a good guess at this eigenvector.

$$\sum_{1 \le j \le n, j \ne i} \frac{1}{i \vee j} \cdot \frac{1}{j^{1/2}} = \frac{1}{i} \sum_{j=1}^{i-1} \frac{1}{j^{1/2}} + \sum_{j=i+1}^{n} \frac{1}{j^{3/2}} \le \frac{1}{i} \left(\int_0^i \frac{1}{x^{1/2}} \, dx \right)$$

$$+ \int_i^n \frac{1}{x^{3/2}} \, dx = \frac{1}{i}(2i^{1/2}) + 2(i^{-1/2} - n^{-1/2}) \le \frac{4}{i^{1/2}}$$

This implies $\sum_j i^{1/2} p_{i,j} j^{-1/2} \le 8\delta$, so if we let $b_{n,k}$ be the expected fraction of vertices in clusters of size k in the model on n vertices, and $|\mathcal{C}_i|$ be the size of the cluster $\mathcal{C}_i$ that contains i,

$$\sum_k k b_{n,k} = \frac{1}{n} \sum_{i=1}^{n} E|\mathcal{C}_i| \le \frac{1}{n} \sum_{m=0}^{\infty} \sum_{i,j} p_{i,j}^m$$

$$\le \frac{2}{n} \sum_{m=0}^{\infty} \sum_{i \ge j} i^{1/2} p_{i,j}^m j^{-1/2} \le 2 \sum_{m=0}^{\infty} (8\delta)^m \le \frac{2}{1 - 8\delta}$$

which completes the proof of the lower bound. ∎

Proof of $\delta_c \leq 1/8$. In this case we need to consider the smallest model, so we set:

$$Q(i, j) = \frac{1}{i \vee j} - \frac{1}{n} \quad \text{when } K < i, j \leq n$$

For those who might expect to see some -1's in the denominator, we observe that they can be eliminated by shifting our index set. By the variational characterization of the largest eigenvalue, for any vector v

$$\rho(Q) \geq \left(\sum_{j=K+1}^{n} v_j^2 \right)^{-1} v^T Q v$$

Again we take $v_j = 1/\sqrt{j}$ for $j > K$.

$$v^T Q v > 2 \sum_{i=K+1}^{n} \sum_{j=i+1}^{n} \frac{1}{i^{1/2}} \frac{1}{j^{3/2}} - \frac{1}{n} \left(\sum_{j=K+1}^{n} \frac{1}{j^{1/2}} \right)^2$$

$\sum_{j=K+1}^{n} 1/j^{1/2} \leq \int_K^n x^{-1/2} \, dx \leq 2n^{1/2}$ so the second term is ≥ -4.

$$\sum_{j=i+1}^{n} j^{-3/2} \geq \int_{i+1}^{n} x^{-3/2} \, dx = 2(i+1)^{-1/2} - 2n^{-1/2}$$

so the first sum is

$$\geq 2 \sum_{i=K+1}^{n} 2(i+1)^{-1} - 2i^{-1/2} n^{-1/2} \geq 4 \sum_{i=K+1}^{n} (i+1)^{-1} - 8$$

where in the second step we have reused a step in the bound derived for the second term. Combining our results

$$\rho(Q) \geq \frac{4 \sum_{i=K+1}^{n} (i+1)^{-1} - 12}{\sum_{i=K+1}^{n} i^{-1}}$$

Letting $q(i, j) = 2\delta \left(\frac{1}{i \vee j} - \frac{1}{n} \right)$ for $K < i, j \leq KN \leq n$ and using

$$\sum_{i=K+1}^{KN} i^{-1} \leq \log N \qquad \sum_{i=K+1}^{KN} (i+1)^{-1} \geq \sum_{i=K}^{KN-1} i^{-1} - 2/K \geq \log N - 1$$

so we have $\rho(q) \geq 8\delta(\log N - 4)/\log N$. If $8\delta = (1 + \epsilon)^4 > 1$ and $N = e^{4+(4/\epsilon)}$ we have

$$\rho(q) \geq (1 + \epsilon)^4 \frac{4/\epsilon}{4 + 4/\epsilon} = (1 + \epsilon)^3$$

for all $K \geq 1$ and the desired result will follow from (2.16) in Durrett and Kesten (1990). Consider the q random graph in (K, NK). There are positive constants γ and β so that if $K \geq K_0$ then with probability at least β, K belongs to a component with at least γNK vertices.

Proof of (2.16) of Durrett and Kesten (1990). Let $M = 1 + (1/\epsilon)$, $L = K/M$ and subdivide (K, KN) into intervals $I_m = [K + (m-1)L, K + mL]$ for $1 \le m \le MN$. Now if $x < x' \in I_m$ then $x'/x < 1 + \epsilon$. If $x \in I_m$ and $y \in I_n$ let

$$\bar{p}(x, y) = \frac{2\delta}{(K + mL) \vee (K + nL)} \qquad \blacksquare$$

The multitype branching process in which an individual of type x gives birth to one of type y with probability $\bar{p}(x, y)$ has spectral radius of the mean matrix $\ge (1 + \epsilon)^2$ if K is large. Since the birth rates are constant on each I_m, we can reduce this multitype branching process to one with MN states. $N = e^{3+(3/\epsilon)}$ and $M = 1 + (1/\epsilon)$ are large but fixed, so we can use the theory of multitype branching processes to conclude that the MN state branching process is supercritical. Now until some interval has more than a fraction ϵ of its sites occupied, the percolation process dominates a branching process with spectral radius $1 + \epsilon$, so the percolation process will be terminated by this condition with probability $\ge \beta > 0$. When this occurs we have at least $\epsilon L = \epsilon KN/(MN)$ occupied sites proving the result with

$$\gamma = \epsilon/MN = C\epsilon^2 e^{-4/\epsilon}$$

This is a very tiny bound on the fraction of vertices in the large component, however, recall the Dorogovtsev, Mendes, and Samukhin (2001) result that

$$1 - g(1) \approx c \, \exp(-\pi/\sqrt{8\epsilon}) \qquad \blacksquare$$

7.3 Subcritical Estimates

To investigate more refined properties, we will confine our attention for most of the rest of the section to model #3 in which an edge from x to y is open with probability $h(x, y) = c/(x \vee y)$. Let $V_{i,j}$ be the expected number of self-avoiding paths from i to j in the random graph.

Lemma 7.3.1. *Suppose $c < 1/4$ and let $r = \sqrt{1 - 4c}/2$. If $1 \le i < j$*

$$P(i \to j) \le EV_{i,j} \le \frac{c}{2r i^{1/2-r} j^{1/2+r}}$$

Proof. Considering all of the self-avoiding paths we have

$$EV_{i,j} = \sum_{m=0}^{\infty} \sum_{*} h(i, z_1) h(z_1, z_2) \cdots h(z_{m-1}, z_m) h(z_m, j)$$

where the starred sum is over all sequences $z_1, \ldots z_m$ of integers in $\{1, \ldots n\}$ so that $i, z_1, \ldots z_m, j$ are distinct. To begin to bound the sum we note that since

$h(x, y)$ is decreasing in each variable

$$EV_{i,j} \leq \sum_{m=0}^{\infty} \int_0^n dx_1 \cdots \int_0^n dx_m \, h(i, x_1)h(x_1, x_2) \cdots h(x_{m-1}, x_m)h(x_m, j)$$

Changing variables $x_i = e^{y_i}$, $dx_i = e^{y_i} \, dy_i$ the above

$$\leq \sum_{m=0}^{\infty} \int_{-\infty}^{\log n} dy_1 \cdots \int_{-\infty}^{\log n} dy_m \, h(i, e^{y_1})e^{y_1}h(e^{y_1}, e^{y_2})e^{y_2} \cdots h(e^{y_{m-1}}, e^{y_m})e^{y_m}h(e^{y_m}, j)$$

The last formula motivates the introduction of

$$p(x, y) = \begin{cases} \frac{1}{4}e^{(x-y)/2} & x \leq y \\ \frac{1}{4}e^{(y-x)/2} & x \geq y \end{cases}$$

which is the transition probability of a random walk with a bilateral exponential jump distribution, and has

$$e^{x/2}h(e^x, e^y)e^{y/2} = 4cp(x, y)$$

that is, on each step the walk is killed with probability $1 - 4c$. Using this notation we have

$$EV_{i,j} \leq \sum_{m=0}^{\infty} \int_{-\infty}^{\log n} dy_1 \cdots \int_{-\infty}^{\log n} dy_m \, \frac{(4c)^{m+1}}{\sqrt{i}} p(\log i, y_1) p(y_1, y_2) \cdots p(y_m, \log j)\frac{1}{\sqrt{j}}$$

$$\leq \frac{1}{\sqrt{ij}} \sum_{n=1}^{\infty} (4c)^n p^n(\log i, \log j)$$

To evaluate the last sum we begin by changing variables $x = y/\lambda$ and using a characteristic function identity from my favorite graduate probability book, Durrett (2004, p. 97).

$$\int e^{itx}\frac{\lambda}{2}e^{-\lambda|x|} \, dx = \int e^{ity/\lambda}\frac{1}{2}e^{-|y|} \, dy = \frac{1}{1 + t^2/\lambda^2} \tag{7.3.1}$$

Using this with $\lambda = 1/2$, it follows that

$$\int e^{itx} \sum_{n=1}^{\infty} (4c)^n p^n(0, x) \, dx = \sum_{n=1}^{\infty} \left(\frac{4c}{1 + 4t^2}\right)^n = \frac{4c/(1 + 4t^2)}{1 - 4c/(1 + 4t^2)}$$

$$= \frac{4c}{1 - 4c + 4t^2} = \frac{4c}{1 - 4c} \cdot \frac{1}{1 + 4t^2/(1 - 4c)}$$

Using (7.3.1) again with $\lambda = \sqrt{1 - 4c}/2$ we conclude that

$$\sum_{n=1}^{\infty} (4c)^n p^n(0, x) = \frac{4c}{1 - 4c} \cdot \frac{\sqrt{1 - 4c}}{4} e^{-|x|\sqrt{1-4c}/2}$$

as it follows that if $i < j$

$$EV_{i,j} \leq \frac{1}{\sqrt{ij}} \frac{c}{\sqrt{1-4c}} e^{-(\log j - \log i)\sqrt{1-4c}/2} = \frac{c}{\sqrt{1-4c}} \cdot \frac{1}{i^{1/2-r}j^{1/2+r}}$$

which is the desired result. ∎

Lemma 7.3.1 is the result we need for the next section, so the reader who is eager to see that proof can go there now. Our next topic is the size of a component containing a randomly chosen site. Dorogovtsev, Mendes, and Samukhin (2001) studied the preferential attachment model in which one new vertex and an average of δ edges were added at each time and the probability of an edge from i to j is proportional to $(d_i + a)(d_j + a)$ where d_k is the degree of k. The CHKNS model arises as the limit $a \to \infty$. Taking this limit of the Dorogovstev, Mendes, and Samukhin (2001) results suggests that the probability a randomly chosen vertex belongs to a cluster of size k has

$$b_k \sim \frac{2}{k^2 \log^2 k} \quad \text{if } \delta = 1/8. \tag{7.3.2}$$

In the subcritical regime one has (see their (B16) and (B17) and not (21) which is wrong)

$$b_k \sim C_\delta k^{-2/(1-\sqrt{1-8\delta})} \quad \text{if } \delta < 1/8. \tag{7.3.3}$$

As the next result shows, once again the physicists are right.

Theorem 7.3.2. *The formulas for b_k hold for model #0 and #1.*

The choice of model here is dictated by the fact that we will use the generating function equation.

Proof. As in the first steps of the proof of (7.1.1), we let $u(y) = 1 - g(1-y)$ and define $v(y)$ by $u(y) = y(u'(0) - v(y))$

$$u'(y) = \frac{1}{2\delta(1-y)} \cdot \left(1 - \frac{y}{u(y)}\right)$$

Differentiating the definition of $v(y)$ gives

$$u'(y) = u'(0) - v(y) - yv'(y)$$

Rearranging and using $u(y)/y = u'(0) - v(y)$ we have

$$v'(y) = \frac{u'(0) - v(y)}{y} - \frac{1}{2\delta y(1-y)} \cdot \frac{u'(0) - v(y) - 1}{u'(0) - v(y)}$$

Combining the two fractions over a common denominator

$$v'(y) = \frac{2\delta(1-y)[u'(0)-v(y)]^2 - (u'(0)-v(y)-1)}{2\delta y(1-y)(u'(0)-v(y))}$$

Using $2\delta u'(0)^2 - u'(0) + 1 = 0$ from Lemma 7.1.2 we have

$$v'(y) = \frac{1-4\delta(1-y)u'(0)}{2\delta(1-y)(u'(0)-v(y))} \cdot \frac{v(y)}{y} + \frac{v(y)^2}{(u'(0)-v(y))y} - \frac{u'(0)^2}{(1-y)(u'(0)-v(y))}$$

Shifting part of the first term to the last

$$v'(y) = \frac{1-4\delta u'(0)}{2\delta(1-y)(u'(0)-v(y))} \cdot \frac{v(y)}{y} - \frac{v(y)^2}{(u'(0)-v(y))y} - \frac{2\delta u'(0)^2 - v(y)}{2\delta(1-y)(u'(0)-v(y))}$$

If $\delta = 1/8$, $u'(0) = 1/4\delta$ and the first term vanishes. If we let $w(y) = u'(0)/(-\log y)$ then

$$w'(y) = \frac{u'(0)}{(\log y)^2} \cdot \frac{1}{y} = \frac{w(y)^2}{yu'(0)}$$

As $y \to 0$ the third term converges to $-u'(0)$, so it is easy to argue that

$$v(y) \sim \frac{u'(0)}{-\log y}$$

To recover the asymptotics for b_k from this we use a result of Flajolet and Odlyzko (1990).

Theorem 7.3.3. *If the singular part of the generating function $f(z) = \sum_n f_n z^n$ has*

$$f(z) \sim C(1-z)^\alpha (\log(1/(1-z))) \quad \text{as } z \to 1$$

then the coefficients

$$f_n \sim C\frac{n^{-1-\alpha}}{\Gamma(-\alpha)}(\log n)^\gamma \quad \text{as } n \to \infty$$

Applying this to g we have $\alpha = 1$ and $\gamma = 1$ so we get $b_k \sim 2/k_2 \log k$ which has the same constant as (7.3.2) but a different power of log. We leave it to someone conversant with Lambert functions to sort out the difference.

If $\delta < 1/8$ then as long as $v(y) = o(y)$ the first term is dominant. If $a = (1-4\delta u'(0))/(2\delta u'(0))$ then

$$v'(y) = \frac{av(y)}{y}$$

is solved by $v(y) = Cy^a$. Noting that $(\log v(y))' = v'(y)/v(y)$ it is not hard to show that if $a < 1$ then $\log v(y)/\log y \to a$ as $y \to 0$. We leave it to the reader to improve this to $v(y) \sim C y^a$. Now $a < 1$ exactly when

$$\frac{1}{6\delta} < u'(0) = \frac{1 - \sqrt{1 - 8\delta}}{4\delta}$$

which translates into $\delta > 1/9$. Using Theorem 7.3.3 we have $\alpha = a + 1$, so $b_k \sim C k^{-(a+2)}$ with

$$a + 2 = 1/2\delta u'(0) = 2/(1 - \sqrt{1 - 8\delta})$$

which agrees with (7.3.3).

When $\delta \leq 1/9$, u has two continuous derivatives, so we have to take away more smooth terms to find the singular part. In general if $k < \frac{1}{2\delta u'(0)} - 1 \leq k + 1$ we can write (recall $u(y) = 1 - g(1 - y)$)

$$u(y) = -\sum_{i=1}^{k} c_i(-y)^i + (-y)^k v(y)$$

and analyze $v(y)$ as before. The details are tedious and are omitted. ∎

7.4 Kosterlitz–Thouless Transition

Let $G_n(c)$ be the random graph on n vertices with connection probabilities $p_{i,j} c/(i \vee j)$. In this section we will prove an upper bound on the size of the giant component due to Bollobás, Janson, and Riordan (2005) which shows that its size is infinitely differentiable at the critical value.

Theorem 7.4.1. *If $\eta > 0$ then for small ϵ the expected size of the largest component in $G_n(1/4 + \epsilon)$ is $\leq \exp(-(1 - \eta)/2\sqrt{\epsilon})n$.*

Proof. Let $\epsilon > 0$. Their first step is to write the random graph $G_n(1/4 + \epsilon)$ as an edge disjoint sum of $G_1 = G_n(1/4 - \epsilon)$ and $G_2 = G_n(2\epsilon)$. To do this, we first construct $G_n(1/4 + \epsilon)$, and then flip a coin for each edge with probability $(1/4 - \epsilon)/(1/4 + \epsilon)$ of heads and $2\epsilon/(1/4 + \epsilon)$ of tails, to decide whether to put the edge in graph 1 or graph 2. The presence of edge (i, j) in G_1 is obviously not independent of its presence in G_2, but is independent of the presence of other edges (i', j') in either graph.

Let $\eta > 0$ and set $\rho = \exp(-(1 - \eta/2)/\sqrt{\epsilon})$. We will consider vertices with index $i \leq \rho n$ early and the others late. Let $f(\epsilon) > 0$ be arbitrary but independent of n. The first step is to argue that G_n is unlikely to contain a component with $f(\epsilon)n$ vertices that does not contain any early vertices. Each early vertex has probability $> 1/4n$ of being directly connected to each late vertex. The probability no early vertex makes a connection is

$$\leq \left(1 - \frac{1}{4n}\right)^{\rho n \cdot f(\epsilon)n} \leq e^{-\rho f(\epsilon)n/4}$$

Our task now is to estimate the probability a late vertex is connected to an early vertex. If we can conclude that the expected number is $\leq C\rho^{1/2}n$, then the giant component has size smaller than $2C\rho^{1/2}n$ and the proof is complete. If $c = 1/4 - \epsilon$ then the constant from Lemma 7.3.1 $r = \sqrt{1-4c}/2 = \sqrt{\epsilon}$, while if $c = 2\epsilon$, $r = \sqrt{1/4 - 2\epsilon}$. To have neater formulas below, we let $\delta = \sqrt{\epsilon}$, and $\gamma = 1/2 - \sqrt{1/4 - 2\epsilon} \approx 2\epsilon$. Letting $N_k(i, j)$ be the expected number of self-avoiding paths from i to j in G_k and using Lemma 7.3.1 we have for $i < j$

$$N_1(i, j) \leq (1/8 + o(1))\delta^{-1}i^{-1/2+\delta}j^{-1/2-\delta}$$
$$N_2(i, j) \leq (2 + o(1))\delta^2 i^{-\gamma}j^{-1+\gamma} \tag{7.4.1}$$

where $o(1)$ depends on δ only. If we observe that $i^{-1/2+a}j^{-1/2-a} = i^{-1/2}j^{-1/2}(i/j)^a$ then using $a = \delta$ in the first case and $a = 1/2 - \gamma$ in the second we have for small δ and $i < j$

$$N_1(i, j) + N_2(i, j) \leq \frac{1}{4\delta\sqrt{ij}} \tag{7.4.2}$$

For $i < j$, we want to bound $N_{12}(i, j)$ the expected number of $i - j$ paths in G consisting of an $i - m$ path in G_1 followed by a $m - j$ path in G_2. From (7.4.1) we have

$$N_{12}(i, j) \leq (1/4 + o(1))\delta \left(\sum_{0<m<i} m^{-1/2+\delta}i^{-1/2-\delta}m^{-\gamma}j^{-1+\gamma} \right.$$
$$+ \sum_{i<m<j} i^{-1/2+\delta}m^{-1/2-\delta}m^{-\gamma}j^{-1+\gamma}$$
$$\left. + \sum_{j<m} i^{-1/2+\delta}m^{-1/2-\delta}j^{-\gamma}m^{-1+\gamma} \right)$$

Bounding by integrals, the above is

$$\leq (1/4 + o(1))\delta \left(i^{-1/2-\delta}j^{-1+\gamma} \int_0^i u^{-1/2+\delta-\gamma}\, du \right.$$
$$+ i^{-1/2+\delta}j^{-1+\gamma} \int_i^j u^{-1/2-\delta-\gamma}\, du$$
$$\left. + i^{-1/2+\delta}j^{-\gamma} \int_j^\infty u^{-3/2-\delta+\gamma}\, du \right)$$

Evaluating the integrals gives

$$\leq (1/4 + o(1))\delta \Big(i^{-\gamma} j^{-1+\gamma}/(1/2 + \delta - \gamma)$$
$$+ (i^{-1/2+\delta} j^{-1/2-\delta} - i^{-\gamma} j^{-1+\gamma})/(1/2 - \delta - \gamma)$$
$$+ i^{-1/2+\delta} j^{-1/2-\delta}/(1/2 + \delta - \gamma) \Big)$$

Combining the second and fourth terms and the first and third, and recalling δ and γ are $o(1)$, the above is

$$= (1 + o(1))\delta i^{-1/2+\delta} j^{-1/2-\delta} - (2 + o(1))\delta^2 i^{-\gamma} j^{-1+\gamma}$$
$$\leq (1 + o(1))\delta i^{-1/2+\delta} j^{-1/2-\delta} \leq (1 + o(1))\delta i^{-1/2} j^{-1/2} \qquad (7.4.3)$$

since $(i/j)^{\delta} \leq 1$. A similar calculation gives $N_{21}(i, j) \leq (1 + o(1))\delta i^{-1/2} j^{-1/2}$. Note that our upper bounds on N_{12} and N_{21} are the same and symmetric in i and j.

We will now fix a late vertex a and estimate the probability that a is joined to some early vertex. If it is there is a path $x_0 = a, x_1, \ldots x_r$, in which all x_i with $i < r$ are late vertices then there is some subsequence $a = y_0, y_1, \ldots y_s = x_r$ such that G_1 contains a $y_{2i} - y_{2i+1}$ path for each i, and G_2 contains a $y_{2i+1} - y_{2i+2}$ path, or vice versa. If s is even then the expected number of paths from a that end with the early vertex b is

$$\leq 2 \sum_{\rho n < y_2, y_4, \ldots y_{s-2} \leq n} \prod_{i=0}^{s/2-1} \frac{(1 + o(1))\delta}{\sqrt{y_{2i} y_{2i+2}}}$$
$$= \frac{2}{\sqrt{ab}} \{(1 + o(1))\delta\}^{s/2} \sum_{\rho n < y_2, y_4, \ldots y_{s-2} \leq n} \prod_{i=1}^{s/2-1} y_{2i}^{-1}$$
$$= \frac{2}{\sqrt{ab}} \{(1 + o(1))\delta\}^{s/2} \left(\sum_{\rho n < z \leq n} z^{-1} \right)^{s/2-1}$$
$$\leq \frac{(2 + o(1))\delta}{\sqrt{ab}} \{(1 + o(1))\delta \log(1/\rho)\}^{s/2-1}$$

Notice that when $s = 2$ this is a little worse than (7.4.3).

For odd s there is an extra N_i at the end. Since $N_1 + N_2$ bounded by (7.4.2) so we get

$$\leq \sum_{\rho n < y_2, y_4, \ldots y_{s-1} \leq n} \prod_{i=0}^{(s-3)/2} \frac{(1 + o(1))\delta}{\sqrt{y_{2i} y_{2i+2}}} \frac{1}{4\delta\sqrt{y_{s-1} y_s}}$$
$$\leq \frac{1}{4\delta\sqrt{ab}} \{(1 + o(1))\delta\}^{(s-1)/2} \left(\sum_{\rho n < z \leq n} z^{-1} \right)^{(s-1)/2}$$
$$\leq \frac{1}{4\delta\sqrt{ab}} \{(1 + o(1))\delta \log(1/\rho)\}^{(s-1)/2}$$

Note that when $s = 1$ this is the same as (7.4.2).

Our choice of $\rho = \exp(-(1 - \eta/2)/\delta)$ implies that

$$(1 + o(1))\delta \log(1/\rho) = (1 + o(1))(1 - \eta/2) \leq (1 - \eta/3)$$

for small δ. Hence when we sum the last two bounds we obtain for small δ

$$\leq \frac{(2 + o(1))\delta}{\sqrt{ab}} 3\eta^{-1} + \frac{1}{48\sqrt{ab}} 3\eta^{-1} \leq \frac{1}{\delta\eta\sqrt{ab}}$$

Summing over b, and comparing with an integral, the probability that a is connected to some early vertex is

$$\leq \sum_{b=1}^{\rho n} \frac{1}{\delta\eta\sqrt{ab}} \leq \frac{2}{\delta\eta} \sqrt{\frac{\rho n}{a}}$$

Summing over a the expected number of later vertices joined to early vertices is

$$\leq \sum_{a=\rho n+1}^{n} \frac{2}{\delta\eta} \sqrt{\frac{\rho n}{a}} \leq \frac{4\rho^{1/2}}{\delta\eta} n$$

Adding in the ρn early vertices the upper bound on the expected size of the largest component is

$$\leq \frac{4\rho^{1/2}}{\delta\eta} n + \rho n \leq \frac{5\rho^{1/2}}{\delta\eta} n$$
$$= \frac{5}{\delta\eta} \exp(-(1 - \eta/2)/2\delta)n \leq \exp(-(1 - \eta)/2\sqrt{\epsilon})$$

when $\delta = \sqrt{\epsilon}$ is small enough.   ∎

7.5 Results at the Critical Value

We turn now to an analysis of model #3 at the critical value. Yu Zhang (1991) studied the percolation process with $p_{i,j} = (1/4)/(i \vee j)$ on $\{1, 2, \ldots\}$ in his Ph.D. thesis at Cornell written under the direction of Harry Kesten, and proved:

Theorem 7.5.1. *If $i < j$ and $i \geq \log^{6+\delta} j$ then*

$$\frac{c_1 \log(i + 1)}{\sqrt{ij}} \leq P(i \to j) \leq \frac{c_2 \log(i + 1)}{\sqrt{ij}}$$

By adapting Zhang's method we can prove similar results for model #3. The upper bound is clean and simple.

Theorem 7.5.2. *If $i < j$ then*

$$P(i \to j) \le \frac{3}{8}\Gamma_{i,j}^n \quad where \quad \Gamma_{i,j}^n = \frac{(\log i + 2)(\log n - \log j + 2)}{(\log n + 4)}.$$

From the upper bound in Theorem 7.5.2 and some routine summation it follows that

$$\frac{1}{n}\sum_{i=1}^{n} E|\mathcal{C}_i| \le 2\sum_{i<j} P(i \to j) \le 6$$

This shows that the expected cluster size is finite at the critical value. This upper bound is only 3 times the exact value of 2 given in Claim 7.1.3. Durrett (2003) has an ugly lower bound. It is an interesting question whether one can prove $c\Gamma_{i,j}^n$ as a lower bound.

Proof. The expected number of self-avoiding paths from i to j is

$$EV_{i,j} = \sum_{m=0}^{\infty}\sum_{*} h(i, z_1)h(z_1, z_2)\cdots h(z_m, j)$$

where $h(x, y) = (1/4)/(x \vee y)$ and the starred sum is over all self-avoiding paths. The sum restricted to paths with all $z_i \ge 2$ has

$$\Sigma_{i,j}^2 \le \sum_{m=0}^{\infty}\int_1^n dx_1 \cdots \int_1^n dx_m h(i, x_1)h(x_1, x_2)\cdots h(x_m, j)$$

Introducing

$$\pi(u, v) = e^{u/2}h(e^u, e^v)e^{v/2} = \begin{cases} (1/4)e^{(u-v)/2} & u \le v \\ (1/4)e^{(v-u)/2} & u \ge v \end{cases}$$

and setting $\log x_i = y_i$, $dx_i = e^{y_i}\,dy_i$ we have

$$\Sigma_{i,j}^2 \le \frac{1}{\sqrt{ij}}G_{0,\log n}(\log i, \log j)$$

where G is the Green's function for the bilateral exponential random walk killed when it exits $[0, \log n]$.

   Suppose the jump distribution is $(\lambda/2)e^{-\lambda|z|}$. Since boundary overshoots are exponential, a standard martingale calculation applied at the exit time from (u, v) shows

$$P_x(T_{(-\infty,u]} < T_{[v,\infty)}) = \frac{(v + 1/\lambda) - x}{(v + 1/\lambda) - (u - 1/\lambda)}$$

the exit probability for Brownian motion from the interval $(u - 1/\lambda, v + 1/\lambda)$. Using this formula and standard reasoning about hitting times, one can show that for the case $\lambda = 1/2$,

$$
G_{K,L}(x, z) = \begin{cases} \frac{1}{4} \cdot \frac{(L-x+2)(z-K+2)}{L-K+4} & z \leq x \\ \frac{1}{4} \cdot \frac{(L-z+2)(x-K+2)}{L-K+4} & z \geq x \end{cases}
$$

To see this is reasonable, note that if we discard the $+2$'s and $+4$'s this is exactly the formula for the Green's function of $\sqrt{8}B_t$.

Taking $x = \log i$, $z = \log j$, $K = 0$, and $L = \log n$ we have for $i < j$

$$
\Sigma_{i,j}^2 \leq \frac{1}{4\sqrt{ij}} \frac{(\log i + 2)(\log n - \log j + 2)}{\log n + 4}
$$

To bound the paths that visit 1 we use

$$
\Sigma_{i,j}^1 \leq \Sigma_{i,1}^2 \cdot \Sigma_{1,j}^2 \leq \frac{1}{16} \cdot 2 \frac{(\log n - \log i + 2)}{\log n + 4} \cdot \frac{2(\log n - \log j + 2)}{\log n + 4}
$$

$$
\leq \frac{1}{8} \cdot 1 \cdot \frac{(2 + \log i)(\log n - \log j + 2)}{\log n + 4}
$$

the upper bound follows. ∎

Printed in the United States
By Bookmasters